A. W. GREELY

HANDBOOK OF ARCTIC DISCOVERIES

Elibron Classics
www.elibron.com

Elibron Classics series.

ISBN 1-4021-4107-6 (paperback)
ISBN 1-4212-7519-8 (hardcover)

This Elibron Classics Replica Edition is an unabridged facsimile of the edition published in 1896 by Sampson Low, Marston and Company, London.

COLUMBIAN KNOWLEDGE SERIES

III

Handbook of Arctic Discoveries

GEN. A. W. GREELY, U.S. ARMY

(Gold Medallist of the Royal Geographical Society, and of the Société de Géographie de Paris)

Knowledge the wing wherewith we fly to heaven.
SHAKESPEARE

COLUMBIAN KNOWLEDGE SERIES
Edited by Professor TODD

NUMBER III

Handbook of Arctic Discoveries

BY

A. W. GREELY
BRIGADIER-GENERAL UNITED STATES ARMY
Chief Signal Officer of the Army

LONDON
SAMPSON LOW, MARSTON, AND COMPANY
M DCCC XCVI

University Press:
JOHN WILSON AND SON, CAMBRIDGE, U.S.A.

EDITOR'S PREFACE

EXPLORATION of the Arctic world has fitly illustrated the truth of BENJAMIN FRANKLIN'S aphorism, that 'an investment in knowledge always pays the best interest;' for during the past two centuries that dreary and seemingly unproductive region of our globe has yielded commercial materials far exceeding a thousand million dollars in value.

This fact alone is sufficient answer to the utilitarian objection to Arctic research; and accurate knowledge of Arctic lands is well worth the having, for purely practical purposes if no other, because the available wealth of this northern world is by no means exhausted. It was commercial interest, indeed, which gave the inception to Arctic investigation, in the search for a northwest passage from Europe to the Orient; and the progress of this search alone led to most important geographic discoveries.

But while Arctic voyagers have penetrated to latitude 83° 24′, within about 450 miles of the geographic pole, an area of the north polar regions exceeding three million square miles remains unexplored. Still it is far from generally known that

Arctic journeys for the purpose of reaching the north pole have been the exception rather than the rule, — a fact which General GREELY distinctly emphasizes. Neither the geographic features of Arctic work, however, nor development along commercial lines, can be said to engross human interest to-day. It is rather that newest phase of polar investigation involved in the increase of scientific knowledge, which recently culminated in establishing the International Circumpolar stations. Only since that event has scientific research in Arctic lands been held at anything like its full value; and it is now clear that this imperfectly known northern world is really a most prominent factor in solving useful problems in the physics of the globe.

Our meteorology; the perfection of theories of the earth's magnetism, requisite in the conduct of surveys and in the navigation of ships; the origin and development of terrestrial fauna and flora; the behavior of ocean currents, — these are fields of practical investigation in which the phenomena of the Arctic world play a very significant rôle. And a knowledge of these phenomena, as yet by no means thorough, is a prime essential to that complete unfolding of Nature, her laws and processes, which is the ultimate aim of scientific investigation.

With the full awakening to these truths for our own planet, not a little curious is the coincidence that tidings similar and almost unmistakable have come from a neighbor world in space, — a world evidently far advanced beyond us in the scale of planetary life. But notwithstanding that disparity, the polar caps on Mars, waxing and waning as they do with the progress of its seasons, appear to afford the true key to the physiographic situation on that planet.

In the interest, then, of farther development in all these lines of human enquiry, — geographic, commercial, and scientific, — I have asked General GREELY to write this *Handbook of Arctic Discoveries.* Without it, if one attempts even a general study of the great problems of the 'Frozen North,' the prodigious array of Arctic literature shows him at once the bald impossibility of the task. So the reader who may desire to take up these enquiries one at a time will not fail to appreciate the obvious advantage of the author's arrangement of his chapters, — topical, not chronological, — while the ample bibliographic additions point in the direction of farther and more specific information.

Mere narrative has been accounted of minimum weight; and compactness has of course necessitated a rigorous condensation, enforcing the omission of many enlivening incidents, but in no sense diminishing the value of the work as a handybook.

DAVID P. TODD.

AMHERST COLLEGE OBSERVATORY,
November 1895.

PREFATORY NOTE

THIS volume represents more than 50,000 pages of original narrative, from which the author has faithfully endeavored to compile such data of accomplished results as may subserve the inquiries of the busy man who often wishes to know what, when, and where, rather than how.

It is confidently believed that no important Arctic geographic addition to knowledge has been omitted from this record. While the original scope of the volume did not include scientific research, yet the more important investigations are alluded to, and the sources whence farther information can be drawn have been indicated in the final chapter.

It should be borne in mind that this is a 'Handbook,' not a Narrative, of 'Arctic Discoveries'; and if the story of adventure is wished, it must be sought in the original special volumes indicated, in the happy bibliographical manner of the *Columbian Knowledge Series*, at the end of each chapter.

If the author has dwelt somewhat fully on American expeditions, and especially on the work he was personally associated with, such a course has been followed with some misgivings, and at the urgent request of the editor of the Series. For this only the indulgence of the public and of the critic are craved.

A. W. GREELY.

WASHINGTON CITY, *November* 1895.

CONTENTS

MAPS

HANDBOOK

OF

ARCTIC DISCOVERIES

CHAPTER I

THE SCOPE AND VALUE OF ARCTIC EXPLORATION

THIS record sets forth rather what men have done than how they have done it, so that such pictures of Arctic travel as here appear are subordinate and incidental to the main subject, — that of Polar Discoveries. If one would gain an adequate idea of the true aspects of such voyaging he must turn to the original journals, penned in the great White North by brave men whose 'purpose held to sail beyond the sunset.'

In those volumes will be found tales of ships beset not only months, but years; of ice-packs and ice-fields of extent, thickness, and mass so enormous that description conveys no just idea; of boat-journeys where constant watchfulness alone prevented instant death by drifting bergs or commingling ice-floes; of land-marches when exhausted humanity staggered along, leaving traces of blood on snow or rock; of sledge-journeys over chaotic masses of ice, when humble heroes straining at the drag-ropes struggled on because the failure of one compromised the safety of all; of solitude and monotony, terrible in the weeks of constant polar sunlight but almost unsettling the reason in the months of continuous Arctic darkness; of silence awful at all times, but made yet more startling by astounding phenomena that appeal noiselessly to the eye; of darkness so continuous and intense that the disturbed mind is driven to wonder whether the ordinary course of nature will bring back the sun or whether the world has been cast out of its orbit in the planetary universe into new conditions; of cold so intense that any exposure is followed by instant freezing; of monotonous surroundings that threaten with time to unbalance the reason; of deprivations wasting the body and so impairing the mind; of failure in all things, not only of food, fuel, clothing, and shelter, — for Arctic service foreshadows such contingencies, — but the bitter failure of plans and aspirations, which brings almost inevitably despair in its train.

Failure of all things, did I say? Nay, failure, be it admitted, of all the physical accessories of conceived and accomplished action, but not failure in the higher and more essential attributes, — not of the mental and moral qualities that are the foundation of fortitude, fidelity, and honor. Failures in this latter respect have been so rare in Arctic service as to justly make each offender a byword and scorn to his fellow-laborers and successors.

Patience, courage, fortitude, foresight, self-reliance, helpfulness, — these grand characteristics of developed humanity everywhere, but which we are inclined to claim as especial endowments of the Teutonic races, — find ample expression in the detailed history of Arctic exploration. If one seeks to learn to what extent man's determination and effort dominate even the most adverse environment, the simple narratives of Arctic exploration will not fail to furnish striking examples.

This volume attempts to justify its title as a 'Handbook of Arctic Discoveries,' whereto one may turn with confidence for succinct accounts of all important voyages and of such minor explorations as are of popular interest. What should appear in, and what should be omitted from, such a volume is obviously a matter of personal opinion, which with the author is the result of diligent research for many years. Under the above plan only such voyages or parts of voyages are treated, as either contributed materially to original geographic knowledge, to the development of other sciences, or as by purpose and action especially merit recognition. Scores of geographic expeditions, following well-known routes, have failed to enhance the knowledge of the world beyond local and minor points, which alter inappreciably charts of normal scale. Original extensions of knowledge — however small or in whatever direction — are always valuable to the world, but supplementary information commands attention only when material and important.

As a work of this kind is necessarily a summary of many narratives, it is proper to here add that references have always been made to original accounts in English or French, otherwise to the most complete translation into one of these tongues.

In preparing this summary of such Arctic explorations as are of historical value, it has been thought that clearness of statement and convenience of reference would be subserved by avoiding the usual method of strict chronological sequence, and adopting a plan whereby all discoveries should be co-ordinated with the particular phase of polar work to which they pertain. Consequently there appear under separate and comprehensive chapters such important and interesting topics as the Northwest Passage, the Northeast Passage, the Franklin Search, International Polar Stations, and Voyages to the North Pole.

This course will tend to dissipate the wide-spread impression that all Arctic voyages have been made for practically the same general purpose, whereas polar research has passed through three distinctive phases: first, for strictly commercial purposes in connection with trade to the Indies; second, for advancement of geographic knowledge; and third, for scientific investigations connected with the physical sciences.

Under such a plan it is not the purpose of the author to dwell on the very early voyages, prior to the sixteenth century, since their extent and results are very largely indeterminate and disputed. Whatever of temporary success marked the daring voyages of the Norsemen and other sea-faring races, yet no permanent settlement, or indeed definite geographical information, remain to the world from any Arctic journey, save in Norway, made prior to the middle of the sixteenth century.

Commercial interests dictated the grand series of voyages wherein England, competing with Spain from the period of the ventures of the CABOTS to the discoveries of BAFFIN, sought for a short route to the Indies across the the Pole or by a Northwest Passage. As the futility of efforts by these routes became more or less apparent, — and as the naval strength of Spain and Portugal ensured their continued monopoly of the growing and valuable trade of the Orient, — the attention of England was turned in sheer desperation to the Northeast Passage, as possibly offering a competing route. While this quest proved impracticable for the sailing ships of the sixteenth century, yet its prosecution inured to the great financial advantage of England, through the establishment thereby of intimate and exclusive commercial relations with the growing and hitherto inaccessible empire of Russia.

The renewal of the true spirit of geographical exploration, in the early part of the present century, gave rise to a series of unparalleled voyages in search of the Northwest Passage, which resulted in the most splendid geographic achievements of the century. These voyages were not splendid alone from the definite results attained, nor from the almost superhuman efforts that ensured success, but also from the lofty spirit of endeavor and adventure that inspired the actors. The men who strove therein were lured by no hope of gain, influenced by no spirit of conquest, but were moved solely by the belief that man should know even the most desolate regions of his abiding place, the earth, and the determination that the Anglo-Saxon should do his part. FRANKLIN said: —

'Arctic discovery has been fostered from motives as disinterested as they are enlightened; not from any prospect of immediate benefit, but from a steady view to the acquirement of useful knowledge and the extension of the bounds of science; and its contributions to natural history and science have excited a general interest. The loss of life in the prosecution of these discoveries does not exceed the average deaths in the same population at home.'

PARRY adds: —

'Such enterprises, so disinterested as well as useful in their object, do honor even when they fail; they cannot but excite the admiration of every liberal mind.'

The latest phase, that of systematic scientific research in the polar regions, has developed almost in its entirety during the past quarter of a century. At present such has become the attitude toward Arctic voyages that an expedition can scarcely command support unless it claims to be in the interests of science. This transition in the status of science relative to Arctic work has been wrought by a few men, and among them there is no other who, by personal effort and argument, has exercised a more potent influence than Baron A. E. VON NORDENSKIÖLD. Doubtless the scientific work of NORDENSKIÖLD incited and gave direction to the well-known efforts of WEYPRECHT that eventuated in the establishment of the International Polar Stations, where geographic and commercial ends were rigidly subordinated to systematic scientific research.

From the voyages under consideration the contributions to material interests and to the sum of human knowledge have been neither scanty nor inconsiderable. The air, the earth, the ocean, even the universe, have disclosed some of their rarest secrets to scientific voyagers in polar lands. Within the Arctic Circle have been located and determined the poles of the triple magnetic forces. In its barometric pressures, with their regular phases, have been found the dominating causes that affect the climates of the northern parts of America, Asia, and Europe. From its sea-soundings, serial temperatures, and hydrographic surveys have been evolved that most satisfactory theory of a vertical interoceanic circulation. A handful of its dried plants enabled a botanist to prophetically forecast the general character of unknown lands, and in its fossil plants another scientist has read unerringly the story of tremendous climatic changes that have metamorphosed the face of the earth. Its peculiar tides have indicated clearly the influence exerted by the stellar worlds on our own, and to its ice-clad lands science inquiringly turns for data to solve the glacial riddles of lower latitudes.

It should not be inferred that more material gains are wanting as the direct result of Arctic research. Although the English navigators failed to reach China and monopolize its trade by a northeast passage, they nevertheless opened a way through the White Sea route to the previously unapproachable empire of Russia, from the Czar of which were obtained large and exclusive trade privileges that inured for many years to the mutual advantage of

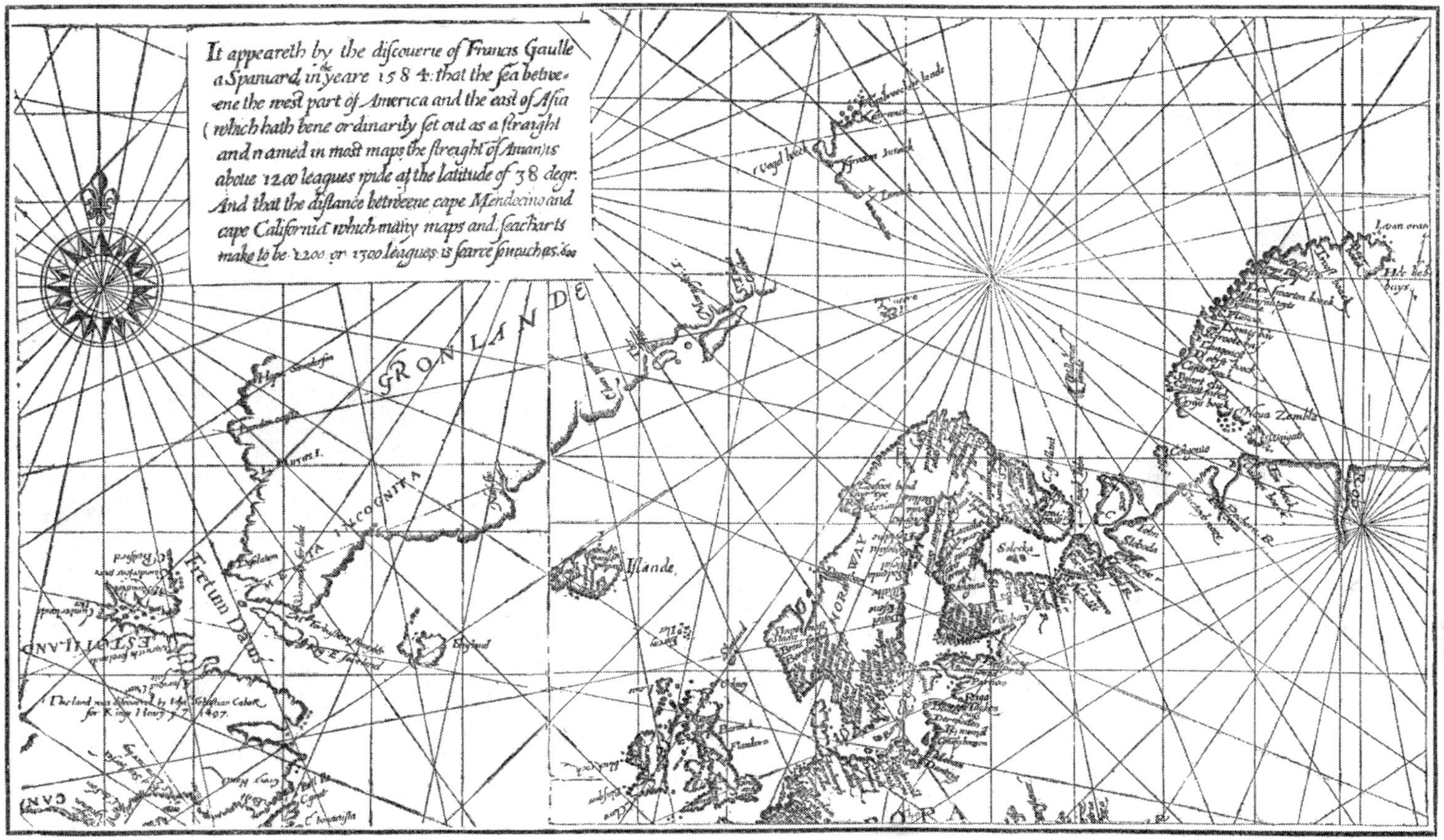

SHAKESPEARE'S NEW MAP. 1600. (After Hakluyt Society Map)

England and Russia. The northern voyages likewise resulted in the valuable whale fishery that, as SCORESBY says, 'in a short time proved the most lucrative and the most important branch of national commerce which had ever been offered to man.' This emphatic statement is devoid of exaggeration in the slightest degree. SCORESBY gives year by year the products of the Dutch whale-fishery in the Arctic seas from 1668 to 1778, which aggregated in value over one hundred millions of dollars. When it is known that SCORESBY himself caught in thirty voyages fish to the value of a million dollars, it will not be considered extravagant to place the products of the British whale-fishery at two hundred and fifty millions of dollars. STARBUCK gives the product of the American whale-fishery from 1804 to 1877 as three hundred and thirty-two million dollars, making the aggregate of the three nations, — America, England, and Holland, — more than six hundred and eighty million dollars. How far this amount should be increased on account of seal, walrus, and other strictly Arctic sea game need not be considered.

BERING failed to outline the definite geographic relations of Asia and America, but his voyage directly resulted in the extremely profitable sea and land fur-trade of the Bering Sea region, and similarly grew up the Hudson Bay trade. Altogether it may be assumed that in a little over two centuries the Arctic regions have furnished to the civilized world products aggregating a thousand millions of dollars in value.

Thoughtful writers have not failed to note the wondrous influence that the initiation of adventurous voyages in the middle of the sixteenth century wrought on the future growth of England. In this respect the consensus of modern opinions justifies the import that NORDENSKIÖLD attaches to the joyous speeding of the Northeast Expedition of WILLOUGHBY and CHANCELLOR, in 1553: 'All was joy and triumph; it seemed as if men foresaw that the greatest maritime power the history of the world can show was that day born.'

The extent of geographic knowledge regarding the Arctic regions at the end of the sixteenth century is shown by the reproduction herewith of the northern part of the chart of 1600. This was called by SHAKESPEARE 'The New Map, with the augmentation of the Indies.' Friesland and Estotiland of the Zeni chart remain, but other traditional lands have given way to the discoveries of Dutch and British navigators, — BARENTS, DAVIS, and FROBISHER. Spitzbergen, Nova Zembla, and Greenland appear, although the last-named country is bisected by Frobisher Strait, which by a natural error (Chapter II) was charted on the wrong side of Davis Strait.

The map of G. DELISLE, in *Voyages au Nord* (Amsterdam 1715), indicates the progress made in the century following the publication of the New Map of 1600. Baffin and Hudson Bays, Jan Mayen and Spitzbergen have been thoroughly explored; Nova Zembla, however, is figured as part of the mainland of Asia, notwithstanding the great Siberian rivers, the Yenisei and Lena, appear, and the coast as far eastward as Swjatoi Noss. Greenland is still bisected by a strait, and the memory of ZENO's Friesland abides off its eastern coast, which shore stretches northward near to the 78th parallel. The marked feature is the wide expanse of unknown region between Asia and America, that remained to be filled in under the inspiration of the voyages of BERING a quarter of a century later, and the entire absence of knowledge regarding the north coast of America.

The map of BARROW, the great British Arctic authority, of 1818, shows that no advance was made in the eighteenth

No. II.

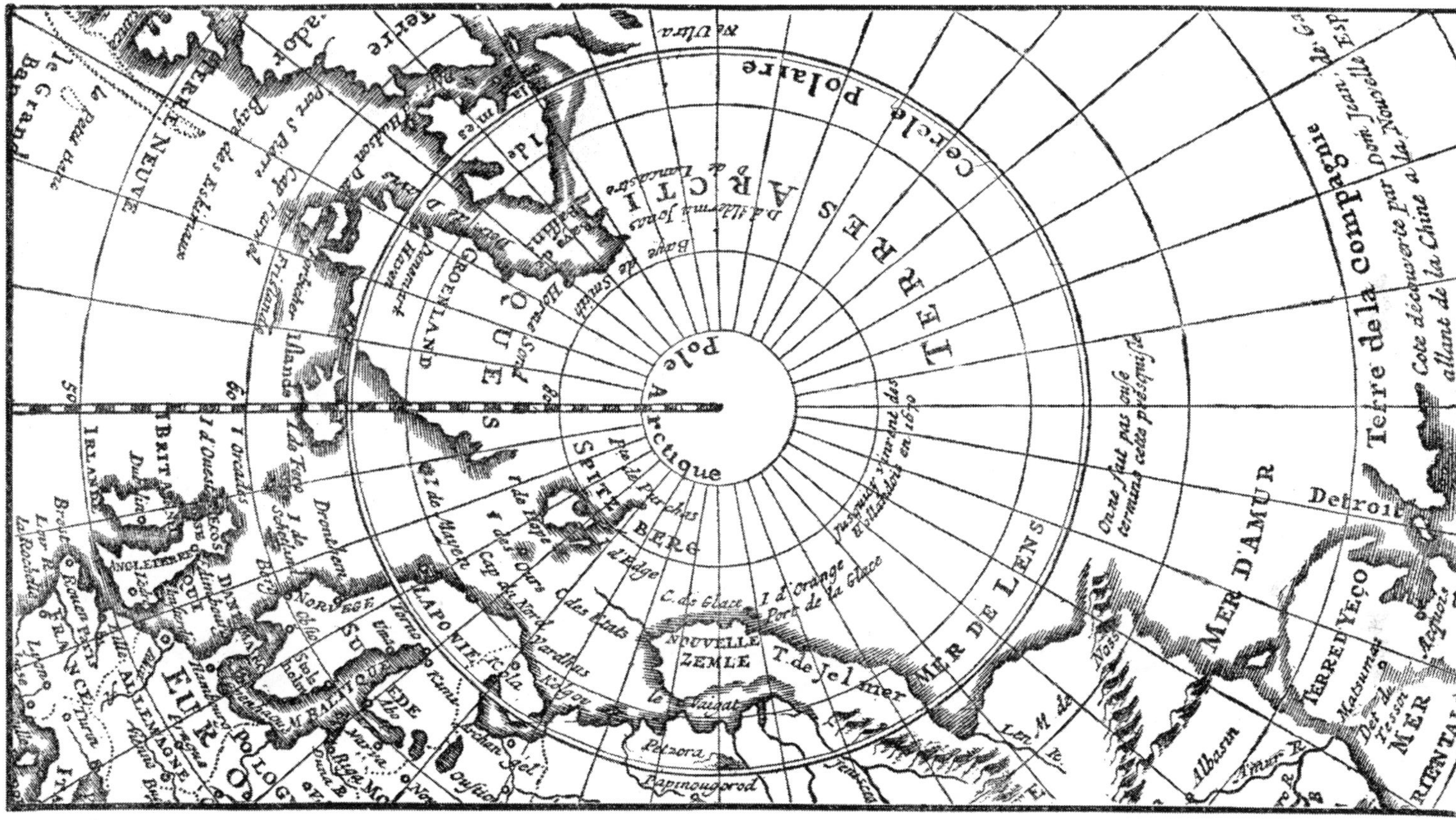

DELISLE'S TERRES ARCTIQUES, 1715

century, and so is not reproduced. In short, it shows retrogression, for not only was the entire coast of North America left vacant, but the discoveries of BAFFIN in 1616 were distrusted, as shown by the omission of Baffin Bay from the chart. BARRINGTON the same year was even more definite, for he entered on his map the legend 'Baffins Bay, .ccording to the relation of W. Baffin in 1616, but not now believed.'

It thus remained for the nineteenth century, with its wealth of industrial inventions and store of indomitable energy, to make the Northwest and Northeast Passages, to outline the northern coast of America, and to discover the archipelagos and islands situated poleward from the three continents of the Northern Hemisphere.

SCORESBY: *The Arctic Regions and Northern Whale-Fishery* (2 v. Edinburgh 1819); RICHARDSON: *Polar Regions* (Edinburgh 1861); MARKHAM (C. R.): *Threshold of Unknown Regions* (London 1876); NORDENSKIÖLD: *Studien u. Forschungen,—Hohen Norden* (Leipzig 1885); HARTWIG: *Polar World* (London 1886); LINDEMANN: *Arktische Fischerei der Deutschen Seestadte,* 1620–1868 (Gotha 1869); RESTE: *Histoire* (ZORDRAGER'S) *des Pêches* (3 v. Paris 1801–2). Read 'Polar Regions' in 9th edition *Encyclopædia Britannica.*

CHAPTER II

EARLY NORTHWEST VOYAGES — TO 1750

THE discovery of the continent of America and the search for a Northwest Passage are inseparably connected, the first event having directly resulted from the latter pursuit. The idea of such a passage originated with JOHN and SEBASTIAN CABOT, father and son.

Sailing, probably in May from Bristol, the CABOTS, 24th June 1497, 'discovered that country which no one before his time had ventured to approach.' The weight of opinion inclines to the belief that this was part of the mainland instead of Newfoundland. If SEBASTIAN CABOT made such claim it was not recognized as a continent by the authorities, as shown by the entry in the Privy Purse accounts of HENRY the Seventh, — 'the 10 of August, 1497. To him that found the New Isle, £10.'

The second voyage under SEBASTIAN CABOT, made with five ships in 1498, was more definite in its results, for during that summer he coasted along the American continent northward to about 67° 30′ N., doubtless discovering the mouth of Hudson Strait, — where appalling dangers and abundant ice obliged him to retrace his way southward, until he reached the vicinity of the thirty-eighth parallel, — still searching for a passage, as far as that part of the firme lande now called Florida.'

Thus from the first venture on the Passage resulted a knowledge of some eighteen hundred miles of American sea-coast. While the explorations thus made disclosed the existence of a great continent as an insuperable barrier to voyagers for China, it incidentally gave such an accurate knowledge of America as led to unexpected advantages in later voyages, enabling explorers to have more specific aims and definite destinations.

CORTEREAL, VERRAZZANO, and GOMEZ (1500–1524) — the last non-English navigators in search of the northwest way to Cathay — failed of their declared intentions, but their voyages were otherwise fruitful. Thence sprang up the lucrative fisheries, pursued by the Portuguese, Basques, and Spaniards in the waters of Newfoundland, and also was developed a more accurate knowledge of the resources of the coast lands of the New World. It is even probable that these adventurous fishermen pushed gradually northward along the rock-bound coast of Labrador and into Hudson Strait and Bay, between 1540 and 1570.

It was only gradually that England came to appreciate the important advantages inuring to Spain and Portugal from fleets manned by danger-hardened sailors, that on the one hand gathered the riches of the northwestern seas, and on the other garnered the wealth of the Orient.

The Northeastern voyages, the founding of the Muscovy Company, and the weakening of the Hanseatic League were preludes to the re-opening of the Northwest Quest, which was to assume practical form after fifteen years of assiduous exertion on the part of Sir MARTIN FROBISHER. ASHER has well said that FROBISHER first gave national character to the search for the Northwest Passage 'by bringing all the most eminent interests in the country — political and aristocratic, scientific and commercial — to bear on this enterprise.' Queen ELIZABETH, Earl WARWICK, LOK, — a wealthy merchant who supplied most of the funds, — and even D^{r} DEE, the official adviser of the closest and most jealous commercial corporation of the day, the Muscovy Company, cordially supported the undertaking. FROBISHER sailed rejoicing, 7th June 1576, with three tiny craft. The pinnace foundered in a violent gale, and his remaining consort deserted, but with stout heart he continued his voyage in the *Gabriel*, — a frail bark of 20 tons burthen, in which to-day the passage of the well-known Atlantic would hardly be attempted. The results of the voyage were a mass of misleading geographic information and quantities of equally worthless gold-earth, which inspired two other disappointing voyages. The gold search ended in financial losses and bitter disappointments, and an excited adventurer transformed an extensive ice-field of Davis Strait into an island, which grew in importance and size as the returning ship sailed toward England. This mythical island, charted as Busse Island, soon disappeared from the maps; but another geographical error, — the Frobisher Strait of southern Greenland, — held its place in Arctic charts for many years, perplexing geographers and explorers, and forming the basis of wild theories, even to this generation three centuries later. FROBISHER'S sailing-chart placed, as did all existing maps, the south end of Greenland in 65° N., and when he reached the mainland of America he fell into the error of believing it to be Greenland, which he naturally supposed to be divided by the newly discovered strait, which was thus mapped as in Greenland instead of in its correct position in America.

One fact was, however, established by FROBISHER'S exploration, on which the supporters of Northwest voyages rested their hopes of ultimate success and arguments for renewed attempts. He had determined the existence of a broad (Frobisher) strait in America, navigable for hundreds of miles between 62° and 63° N., and of a second (Hudson) strait broader and more easily navigated, between 60° and 62° N.

The first great advance toward the discovery of the Northwest Passage resulted from the three voyages of JOHN DAVIS, of Sandridge, a distinguished seaman whose abilities as a sailor were no less conspicuous than his skill as a pilot and knowledge as a navigator.

Under letters patent, the right of search by all northerly routes was granted to a company which sent DAVIS forth in 1585 with the *Sunshine*, 60 tons, and the *Moonshine*, 35 tons. Sighting Greenland 20th July, he well describes it as a 'land being very high and full of mighty mountains all covered with snowe, no viewe of wood, grasse or earth to be seene, and the shore two leagues into the sea full of yce. The loathsome view of the shore and irksome noyse of the yce was such as to bred strange conceites among us.' On 29th July he was off the west coast near Godthaab, having 'past al the yce and found many green and pleasant Isles bordering upon the shore.'

Crossing the strait which now bears his name, DAVIS reached Cape Dyer, sailed nearly to the head of Cumberland Sound, and returned to Dartmouth, 30th September.

DAVIS'S voyage of the following year, made in the same vessels, with the addition of the *Merimade*, 120 tons, and a pinnace of ten tons, added nothing to the Northwest discoveries of the preceding year. However, he stretched southward along the American coast to Labrador without observing the entrance to Hudson Strait. On this voyage the unfortunate little pinnace, *North Star*, was lost in a violent storm.

The third voyage of 1588 was made in the *Elizabeth*, the *Sunshine*, and the *Ellen*. Following the west coast of Greenland to the vicinity of Sanderson's Hope,

DAVIS found the 'sea all open to the westward and northward' at his farthest northing, 72° 12′ N. Thence he steered westward, passing through the dreaded middle ice-pack of Baffin Bay 'a mighty bank of ice,' says DAVIS. From Mount Raleigh, the western side of Davis Strait, he reports that there was no ice toward the north, but a great sea, 'free, large, very salt and blue, and of indeterminable depth.' This last voyage, one of most reckless gallantry, must have taxed to the utmost the skill and seamanship of DAVIS in navigating his fleet of tiny vessels through unknown waters, amidst constantly recurring dangers the more terrific from their novelty. The remarkable discoveries of DAVIS covered the west coast of Greenland from Cape Farewell to Sanderson's Hope, and on the American side, from Cape Dyer, Cumberland Island, to Southern Labrador.

His descriptions of the Greenlanders are quaint, curious, and instructive, showing them to have been, three centuries ago, the same 'tractable people, void of craft or double-dealing' as we know them to-day. Of the voyage of WEYMOUTH in 1602 Fox remarks: 'He neyther discovered anything more than DAVIS, nor had any sight of Greenland; . . . yet these two, DAVIS and he, did, I conceive, light HUDSON into his straits,' into which he sailed 'an hundred leagues west and by south.' The ventures of KNIGHT and HALL, 1605–7, were failures, and the next honors were gained by a man whose recorded life runs only from 1607 to 1611. Four years are a brief period, but it was enough for HENRY HUDSON to make an impress on his own generation of great and adventurous seamen, and yet more to render his name imperishable in history. He visited almost all the known Arctic lands of Europe and America, and in whatever direction he turned his energies, remarkable successes followed.

Early Northwest Voyages — to 1750

HUDSON's earlier voyages, 1607–9, along the coasts of Spitzbergen and Nova Zembla, are elsewhere (Chapters III and V) mentioned. In 1611 he returned to service under England, and in a voyage at the expense of three gentlemen, Sir THOMAS SMITH, Sir DUDLEY DIGGS, and Master JOHN WESTONHOLME, — attempted the search for a Northwest Passage.

HUDSON sailed on his fourth known voyage, in the *Discovery*, from London, 17th April 1810, to try 'if, through any of those inlets which Davis saw . . . on the westerne side of Fretum Davis any passage might be found to the other ocean.'

Touching at Iceland and later making the east coast of Greenland in 65° 30′ N., HUDSON found the shore ice-bound. Standing southward and imagining one of the many inlets to be Frobisher Straits, he rounded Cape Desolation (Farewell), sailed west and entering a (Hudson) strait passed, 21st July, to a point where he 'found the sea more growne than any wee had since wee left England.' He was in Hudson Bay.

CABOT, DAVIS, and WEYMOUTH had visited the mouth of Hudson Strait, and there is the best of cartographic proof that the Portuguese fishermen had even entered the bay. Their charts of 1558 show their familiarity with the strait, while the atlas of ORTELIUS in 1570 definitely outlines the (Hudson) bay. However, HUDSON's achievements were so striking, and his fate so tragic, that his name properly abides with the great inland sea, in which as elsewhere he distanced his predecessors in successful exploration.

His ship beset with ice, his officers mutinous, and his men sick with fear, yet HUDSON continued his voyage, and following the east side of the bay reached its south-eastern extremity, James Bay, where, frozen-in the 12th of November, the expedition wintered. One man died and provisions were scanty, but HUDSON was a man of resources, and so supplemented short rations by hunting and fishing, which furnished forth large amounts of suitable food. During the autumn he had displaced his mate and boatswain for mutinous conduct, and the winter bred bad feeling, which culminated after HUDSON had broken out of winter quarters and was exploring the western shore. The mutineers, led by HENRY GREEN, a protégé who owed everything to the generosity of HUDSON, set adrift in a small shallop, 21st June 1613, the captain, five sick men, and JOHN HUDSON, possibly HUDSON's son. A ninth man, JOHN KING, the carpenter, loyally cast in his fate with his commander. Thus perished by the basest ingratitude HENRY HUDSON, for diligent search the following year by Sir THOMAS BUTTON was fruitless. The mutineers fared little better, for GREEN and four others perished at the hands of savages, and the rest barely escaped from death by starvation in their wretched homeward voyage to merited imprisonment in England.

HUDSON died living up to his noble rule that men should 'achieve what they had undertaken, or else give reason wherefore it will not be.' If he found not the Northwest Passage, his discoveries in connection with the search therefor gave to England the wealth of furry land-game in the Hudson Bay territory as a supplement to the inheritance of finny sea-game in the Spitzbergen whale-fishery, the result of his earlier voyages.

Sir THOMAS BUTTON, pursuing the route laid down by his unfortunate predecessor, also found Hudson Bay full of dangers. He lost five men by the natives, and quite a number died at Port Nelson, where he wintered. His voyages gave no information as to HUDSON's exact fate, but BUTTON crossed the bay from east to west, — first of navigators, — and explored its western shores.

His discoveries include Nelson River, Button Bay, Button Island, Ne Plus Ultra (Roe Welcome) Strait, and Southampton Island. Unfortunately for his successors, he believed that the passage to the South Sea could be found out of the northwest part of the bay, and it was a century before this idea could be dispelled completely — by PARRY. BYLOT, in 1615, confirmed BUTTON'S discoveries, but added nothing to them.

Emulous of English discoveries, enterprising Denmark sent JENS MUNK on the search. Entering Hudson Bay there is good reason to believe that he made discoveries along the western shore. He wintered, not in Chesterfield inlet, but in Churchill River, where his misfortunes were so great as to destroy his credit. Disease carried away his men almost to a man, — some accounts say 62 out of 64 seamen, — and with untold difficulties MUNK with his few remaining men made their return voyage in the smaller of his two vessels.

JAMES and so-called 'Northwest' Fox, sailed in 1631, the latter bearing a letter from the King of England to the Emperor of Japan. JAMES had, as he styles it, a dangerous and lamentable voyage, losing several men, his ship, and later his house by fire at Charlton Island, in the southern extremity of Hudson Bay, now called James Bay. No discoveries resulted from his voyage.

It is doubtful if the region covered by Fox was new; but he affixed his name to a strait, — Fox Channel, where he reached 66° 50′ N. 77° W., 21st September 1631, — and named an (Southampton) island Sir Thomas Roe's Welcome. What he failed to obtain in reputation from the actual work done, he gained from a facile pen in his quaint and somewhat humorous account of this voyage, — styled 'Northwest Foxe.' A few islands, — Brooke, Cobham, and Brigg's Mathamatick's, as he calls them, appear to cover his original work.

To close with efforts by Hudson Bay, it may be said that the voyage of MIDDLETON in 1741 resulted in the discovery of Wagner River and Repulse Bay, thus affording convincing proof of the impossibility of passing to the west from Hudson Bay. Sir ARTHUR DOBBS unjustly questioned the accuracy of MIDDLETON'S work and claimed that Wagner River was a strait leading to the South Sea. A second expedition, — the *Dobbs* and *California,* under MOORE and SMITH, — sent out in 1746, through DOBBS' influence, reported to his discomfiture that the Wagner was a river.

Returning to the only route by which success was possible, DAVIS was there followed by another able seaman and great discoverer, WILLIAM BAFFIN, who, in the *Discovery,* a craft of only 55 tons, sailed 16th March 1616, from Gravesend. He sighted Greenland 14th May, and on the 30th of that month had reached DAVIS' farthest point, Sanderson's Hope, in 72° 41′ N. On 9th June, he was stopped by ice at Baffin Islands, 73° 54′ N. Leaving his anchorage 18th June, he took what is known as the 'Middle Passage' across Melville Bay, and reached, 1st July, an open sea, — the 'North Water' of the whalers of to-day. Passing Capes York, Atholl, and Parry, he yet pushed northward, and on 5th July attained his farthest point within sight of Cape Alexander. His latitude, about 77° 45′ N., remained unequalled in that sea for two hundred and thirty-six years. BAFFIN in quaint language says he was forced by ice 'to stand backe some eight leagues to an island we called Hakluits Ile — it lyeth betweene two great Sounds, the one Whale Sound, and the other Sir Thomas Smith's Sound; this last runneth to the north of 78, and is admirable in one respect, because in it is the greatest variation of the compasse of the world known; for by divers good observations I found it to be above five points, or fifty-six degrees varied to the westward.'

A few days later BAFFIN turned southward, having in this wonderful voyage sailed over three hundred miles farther north than his predecessor, DAVIS. He thus added to geographical knowledge Ellesmere and Prudhoe Lands, and Baffin Bay, with its radiating sounds of Smith, Jones, and Lancaster. With this voyage ended all efforts to discover a route to Cathay and the Indies by Davis Strait; for two centuries the waters first navigated by BAFFIN remained unvexed by any keel, and the very credit of his discoveries passed from the mind of man.

See Hakluyt Society Publications, London, v. d.: MAJOR: *Voyages of the 'Zeni'* (1873); RUNDALL: *Voyages towards the North-West*, 1496–1631 (1859); COLLINSON: *Three Voyages of Martin Frobisher* (1867); MARKHAM, A. H.: *Voyages of John Davis* (1880); ASHER: *Henry Hudson the Navigator* (1860); MARKHAM, C. R.: *Voyages of William Baffin*, 1612–22 (1881); BARROW: *Coat's Geography of Hudson Bay* (1852).

GOLDSMID: *Hakluyt's Voyages of English to America before* 1600 (4 v. Edinburgh 1891).

CHAPTER III

NOVA ZEMBLA

AS is well known, the principal part of this land is divided by Matthew Strait (Matotschkin Schar) into two large islands. The name Nova Zembla originally covered the southerly of these twin lands, but it is now properly applied to the whole group.

The islands of Nova Zembla have always been classified as uninhabitable, in common with all other outlying polar lands, except Greenland. For this reason the late and partly successful efforts of the Russians to colonize the southern island give a renewed interest to Nova Zembla.

In this connection quite a number of Samoyed families settled about a dozen years since at Karmakuly, Moller Bay, in the northern part of Gooseland. There the climate is milder than in their former Siberian home, vegetation is abundant for reindeer and other animals, and the transition from summer hunters — migrating with the changing seasons north and south to and from the Siberian mainland — to permanent residents of Nova Zembla, is not so great a change for the Samoyeds as would appear.

It is unknown when the Novgorod hunters first visited Nova Zembla, but it was probably several centuries earlier than the fateful voyage of WILLOUGHBY in 1553, or of his successor STEPHEN BORROUGH in 1556 (Chapter IV). Some claim that WILLOUGHBY sighted Nova Zembla, but the opinion of NORDENSKIÖLD is more probable, that the land seen was Kolgujef Island. This leaves to BORROUGH the honor

No. III.

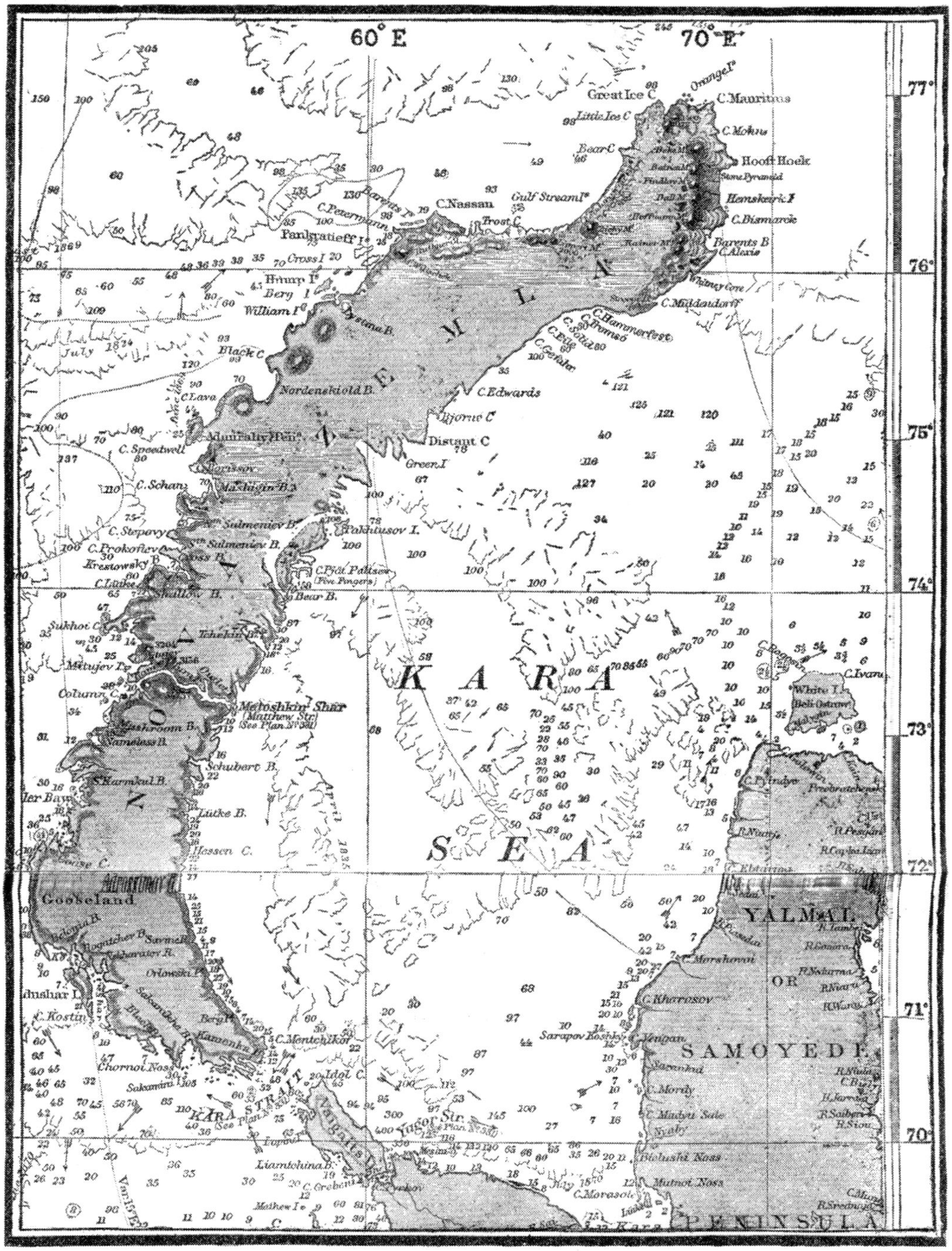

NOVA ZEMBLA (Hydrographic Chart, U. S. Navy)

of being the first European known to visit and give definite information regarding the country. The way was shown, however, by Russian fishermen, one of whom, LOSKAK, called it 'New Land,' or Nova Zembla. Then as now the Samoyeds pastured reindeer on Waigat Island, south of Borrough (or Waigat) Strait, plied their boats to north or south, pitched their conical deerskin tents and set up bloody idols near their sacrificial mounds.

On the adjacent mainland assembled each summer, coming distances of four or five hundred miles, Samoyed hunters and reindeer-owners for pasturage and game, followed closely by Russian traders for barter and speculation. Centuries have seen this routine of coming and going, until of late years, as NORDENSKIÖLD relates, a small permanent village, Chabarova, part Russian, part Samoyed, has grown up on the mainland south of Yugor Strait.

The next knowledge of Nova Zembla resulted from the efforts of the Amsterdammers to reach China by sending a vessel round the north of the island. The duty was entrusted to WILLIAM BARENTS, a man of great perseverance and indomitable courage, a mariner of experience, an observer so skilled that modern observations confirm his positions, and a navigator so capable as to fear comparison with none other. Sailing, 5th June 1594, from Texel in the *Mercury*, 100 tons, BARENTS sighted, 4th July, Langenes Point, Nova Zembla, 73° 25′ N., whence he traced northward the west coast beyond Cape Nassau. Here his progress was impeded by immense quantities of ice, but he finally reached, 31st July, the Orange Island, north of the seventy-seventh parallel. 'Finding,' says DE VEER, 'that he could hardly get through to accomplish and end his intended voyage, his men also . . . would sail no further,' BARENTS returned homeward.

In his twenty-five days' struggle, from Cape Nassau to

Orange Island and back, PETERMANN has shown that BARENTS put his ship about eighty-one times and sailed 1546 miles, a remarkable voyage for a sailing craft of 100 tons. Turning southward, BARENTS joined near Jugor Strait the Dutch vessels that had unsuccessfully attempted the southern route, and with them returned to Holland.

BARENTS sailed the following year, with other Dutch ships, to try the route to Cathay by the Waigat Strait, which point was the farthest reached.

Undismayed by unsuccess, Amsterdam decided to attempt once more the northerly passage, and sent forth, 18th May 1596, two vessels, commanded by JOHN CORNELIUS RYP and JACOB HEEMSKERCK. BARENTS sailed as chief pilot to HEEMSKERCK, but as the master was not a sailor the navigation of the ship necessarily devolved on BARENTS. After the discovery of Spitzbergen (Chapter V) and return to Bear Island, the two ships parted, RYP returning to Spitzbergen, while BARENTS, who had steadily asserted that their course was too far to the west, sailed east to Nova Zembla, which was reached 17th July. Following its western shore they rounded the northeastern extremity without much difficulty. The coast then trending to the south, they had great hopes of complete success, but falling in with heavy and constantly moving ice-packs they were forced, 26th August 1596, into a bay, Ice Haven, on the east coast, from which the ship was destined never to emerge.

A description of this wintering in some detail is not out of place, for the ship, not fitted with a view to such contingency, was entirely unsuited for habitation. Both the rigors of an Arctic night and the diseases incident thereto were unknown. Food was limited in quantity, fuel lacking, and clothing unsuitable. Wisely believing that a hut was their proper means of shelter, they determined on erecting it. Exploring the country 'they found great store of wood . . . trees roots and all, driven upon the shoare, either from Tartaria, Muscovia, or elsewhere, for there was none growing upon that land, wherewith, as if God had purposely sent them to us, we were much comforted.'

This drift-wood, their salvation from death by cold, was eight miles distant from the site of the house. To less energetic men it would have seemed impossible to haul so far timbers for the house, and wood for the winter's fuel. It proved a terrible task, not only in view of the physical labor, the increasing cold, and rapidly shortening days, but also from the fact that bears were so numerous and ferocious that the men were obliged to go armed as they hauled their heavy loads over the rough country.

Their sufferings from cold, confinement, and storm, their danger from bears, their courage, cheerfulness, and faith under such conditions, and the energy, skill, and perseverance with which they met all emergencies are best illustrated by brief quotations from the narrative of DE VEER, one of the party: —

'It blew so hard and snowed so fast that we should haue smothered if we had gone out into the air; and to speake truth, it had not beene possible for any man to haue gone one ship's length, though his life had laine thereon; for it was not possible for us to go out of the house. One of our men made a hole open at one of our doores, . . . but found it so hard wether that he stayed not long, and told us that it had snowed so much that the snow lay higher than our house. . . .

'The beares came fiercely towards us, that had no other armes to defend us withall but onely the two halberds, . . . wee still gaue them worke to do by throwing billets and other things at them, and euery time we threw they ran after them, as a dogge useth to doe at a stone that is cast at him. . . . At the last, as the beares came fiercely upon us, we stroke one of them with a halberd upon the snoute, wherewith shee gaue back when shee felt her selfe hurt, and went away, which the other two yet were not so great as she perceiuing, ran away; and we thanked God that wee were so well deliuered from them. . . .

'It frose so hard that as we put a nayle into our mouths (as when men worke carpenters worke they use to doe), there would ice hang thereon when we tooke it out againe, and made the blood follow. . . .

'It was so extreame cold that the fire almost caste no heate; for as we put our feete to the fire, we burnt our stockings before we could feele the heate. . . . And, which is more, if we had not sooner smelt than felt them, we should haue burnt them quite away ere we had knowne it. . . .

'We alwaies trusted in God that hee would deliuer us from thence towards sommer time either one way or other. . . . We comforted each other, giuing God thanks that the hardest time of the winter was passed, being in good hope that we should liue to talke of those things at home in our owne country.'

On 25th January they were surprised by the return of the sun fourteen days earlier than it was due. This phenomenon has given rise to much discussion as to the cause, but it is probable that it was due to extraordinary atmospheric refraction attendant on extreme cold. The date appears well determined, although, their clock failing, time was kept by a twelve-hour hour-glass, which had to be watched and promptly turned.

With advancing spring it became evident that the harbor ice would not break up early, and near the end of May it was decided that their only hope of escape was a retreat by small boats to Lapland, distant more than a thousand miles. Two men had died, and two others, BARENTS being one, were sick. Finally, the 14th of June, they quit their winter quarters, and such was their sense of duty that they spared a part of room to bring back packets of their most costly merchandise.

The strength of BARENTS was nearly spent, and every day of exposure brought him rapidly to the end. Cold and hunger, exposure and disappointment could impair his physical energies, but they could not break the spirit of the man, who died as he had lived, with his sense of duty and effort unchanged.

Three times in strength and vigor, he had passed the north end of Nova Zembla, and now the fourth time, a dying man, came to it. DE VEER says: 'And being at the Ice Point the maister called to WILLIAM BARENTS to know how he did, and WILLIAM BARENTS made answeare and said, Quite well mate. I still hope to be able to run before we get to Wardhuus. Then he spak to me and said: Gerrit, if we are near the Ice Point, just lift me up again. I must see that Point once more.'

The end came, 20th June 1596, while he was conning and criticising the chart, whereby the boat journey was being made. DE VEER relates, 'WILLIAM BARENTS looked at my little chart, which I had made of our voyage, and we had some discussion about it; at last he laid away the card and spak unto me, saying, Gerrit, give me something to drink and he had no sooner drunke but he was taken with so sodain a qualme, that he turned his eies in his head and died presently. The death of WILLIAM BARENTS put us in no small discomfort, as being the chiefe guide and onely pilot on whom we reposed our selues next under God; but we could not striue against God, and therefore we must be content.'

The character and conduct of BARENTS were such as endeared him equally to master and to men. Of his voyages, BEKE says: 'BARENTS made so many discoveries and traced so large an extent of coast, both of Spitzbergen and Nova Zembla, that the surveys of all of the whole of our recent explorers (1853), put together, are insufficient to identify all the points visited by him.'

While HEEMSKERCK was no seaman by profession, yet he had marine experience, and by no means played a subordinate part in the expedition. All matters were referred to him, and his decision was final. His success in making the boat journey to Kola, after BARENTS' death, indicates that he then had the qualities of daring, decision, and judgment which he so displayed in after years.

As MOTLEY says, 'Incapable of fatigue, of perplexity or of fear,' HEEMSKERCK as a privateersman captured an armed Portuguese vessel of four-fold force. Later, as admiral of the Dutch fleet in 1607, attacking at Gibraltar a Spanish war-fleet of 21 ships, HEEMSKERCK captured, sank, or burned every vessel, and thus ensured to Holland by the south the way to the Indies he had heroically attempted by the north.

With the voyages of BARENTS, advances in geographic knowledge as regards Nova Zembla ceased for nearly two centuries, until the spirit of exploration that imbued Russian sailors in the eighteenth century was turned toward the neglected coasts of this desolate land. It is true that HENRY HUDSON in his third voyage visited the shores of Karmukal Bay, but unfavorable ice conditions prevented his reaching the northerly coast, and his journey to the southward added little to the results of his predecessors. Doubtless also the Dutch whalers of the seventeenth century passed to the northward of Nova Zembla, as have the walrus-hunters of to-day. In 1664 it is said that a whaler, DE VLAMINGH, succeeded in rounding the northeast point of Nova Zembla. However that may have been, it was near two centuries after the voyages of BARENTS before other information of value or interest was acquired. It is said that SAWWA LOSCHKIN visited in 1760 the unknown east coast, expecting to find an abundance of fur animals. The chronicles run that he wintered two years on that coast, attained a latitude of 76° 09′ N., reached the northeastern extremity of the island, and returning by the west shore circumnavigated Nova Zembla. The results of the voyage appear to have been exaggerated, and were not utilized by his map-making compatriots. As Nova Zembla is now known to extend to 77° N., as set down by BARENTS, it could not have been circumnavigated in 76° 09′ N. Farther, the 1864 chart of ERMAN, drawn after Russian data, leaves the east coast very uncertain to the north of 75°, only three Dutch names appearing thereon.

A few years later, 1768–69, the Russian pilot ROSMISLOV explored the region of Matthew Strait, which passage was said to have been long known to local fishermen, and there wintering lost his ship, a leaky, unseaworthy craft.

The efforts of the Russian government during the present century have been productive of much valuable knowledge regarding the greater part of the coasts. From 1821 to 1824, Lieutenant (afterwards Admiral) LÜTKE was engaged in exploring the west coast in a small brig. He twice failed to pass beyond Cape Nassau, his farthest point, but his observations laid down definitely and accurately the more southerly coast-line and its indentations, while his hydrographical contributions are valued both in their local and in their general relations.

Between 1832 and 1835 PACHTUSSOW surveyed the east coast as far north as Pachtussow Island, 74° 24′ N., thus making the most extensive contribution on record to our knowledge of this shore. He wintered twice on the island, explored by sledge and ship, and moreover made a most valuable series of observations, astronomical, geodetic, meteorological, and hydrographical. Few explorers have done more with equal means and opportunities, and his voyages are the most remarkable in the annals of Nova Zembla. He paid for his unremitting exertions by his death, from overwork, at Archangel on his return. ZIVOLKA, who accompanied him on the second voyage, died in an unsuccessful expedition in 1838.

In 1837 the first scientific examination of the natural history of Nova Zembla was made by the well-known naturalist, K. E. VON BAER, accompanied by the geologist LEHMANN, and Lieutenant ZIVOLKA. This voyage, confined to locations previously known, gave BAER great temporary reputation, which his energy and industry merited. He, however, generalized on the natural conditions of the eastern polar sea on scanty data, which led to conclusions not accepted in the light of fuller knowledge. Likewise his geographic elucidations were in most cases theoretical, and resulted in the wide dissemination of errors that have only lately disappeared from our charts. In an address on Nova Zembla discoveries in 1838, BAER practically adopted the Russian map of ZIVOLKA, and rounding of the island a short distance beyond Cape Nassau, omitted the entire northeastern part of Nova Zembla, which neither BAER, ZIVOLKA, nor any other man had seen since the days of BARENTS. It is interesting to note that the land thus erased from the map was that part of Nova Zembla that had been most fully explored, it having been traced by BARENTS four times with such accuracy that recent observations only cause wonder for the correctness of the work with the rude instruments of his day.

The dimunition of northern game in the northwestern

seas has turned in late years the course of hunting to the less accessible waters surrounding Nova Zembla. Among the many successful navigators are CARLSEN, QUALE, ULVE, JOHANNESEN, and PALLISER, the last an English sportsman who sailed half a degree north of Cape Nassau in 1869. ULVE, 1st August 1870, reached 76° 47′ N., 59° 17′ E., some fifty miles northeast of Cape Nassau; and QUALE in 1871 sailed to the east of the Obi, 75° 22′ N., 74° 35′ E.

The most remarkable voyages are those of the Norwegian hunter, E. H. JOHANNESEN, whose hydrographical observations during his successful navigation of Barents and Kara seas in 1869 won him a silver medal, presented by NORDENSKIÖLD in the name of the Swedish Academy of Sciences. The following year, after a most successful hunting trip in Kara Sea, JOHANNESEN with a full ship returned to Norway by the north of Nova Zembla, passing its most northern point 3d September. In 1878, he made a more remarkable voyage. Rounding the north of Nova Zembla 22d July he discovered and circumnavigated in 77° 31′ N., 86° E., a small (Lonely) island snow-free and frequented by bears, birds, and seals.

The voyage of Captain ELLING CARLSEN, in which he circumnavigated Nova Zembla in 1871, is yet the more interesting from the fact that his was the first ship, after an interval of 275 years, to follow BARENTS into the harbor of Ice Haven. CARLSEN, in the sloop *Solid*, rounded Nova Zembla and anchored at Cape Hooft, where he fell in with Captain MACK. Together they carefully determined the longitude of the most northeasterly point of Nova Zembla, which is in 67° 30′ E., instead of 73° E. as located on the sailing charts. They moreover determined the trend of the northeast part of the island, which is quite to the north instead of to the northwest as charted. These observations, it may be added, confirm the accuracy of the old Dutch navigators.

The winter quarters of BARENTS and HEEMSKERCK were visited on the 9th, 12th, and 14th of September, and as many relics as could be conveniently obtained were brought away. The house, about 50 by 30 feet in size, was entirely decayed, but a coating of ice protected many articles, which were recovered in a wonderful state of preservation. These consisted principally of books, engravings, carpenter's-tools, cooking utensils, candle-sticks, navigation instruments, portions of fire-arms, and a clock. Eventually these relics, 78 in number, passed into the possession of the Dutch government, and are now exhibited at the Hague in a model-room, the fac-simile of BARENTS' original house at Ice Haven.

A Norwegian captain, M. GUNDERSEN, visited Ice Haven in 1875, and an English sportsman, C. L. W. GARDNER, in 1876; both brought back other relics, those of GARDNER comprising the record left by BARENTS, which is the only known writing of the great navigator. GARDNER'S observations place the winter-quarters in 76° 12′ N.

The establishment of a permanent Samoyed settlement at Karmakuly, Moller Bay, under Lieutenant TYAGHIN, points to an ultimate knowledge of the interior of Nova Zembla, of which little is known. The location of the International station, 1882–83 (Chapter XVI) on Little Karmakuly island, Moller Bay, not only resulted in the accumulation and discussion of the magnetic, meteorological, and hydrographical conditions of the southern isle, but also of the fauna and flora. The observations were not entirely local, for Dr GRINETSKY in May, 1883, crossed from Little Karmakuly to the Kara Sea, and later visited Matthew Strait.

The latest explorations of Nova Zembla have also been

scientific, under a Russian, M. K. NISSILOF, who has passed three winters, 1887–88 to 1890–91, on the southern island, the last at the western entrance of Matthew Strait. While charting and sounding along the coast three new islands were discovered, one (Possiet) being some twenty by three miles in area. Valuable botanical, mineral, and zoölogical collections were obtained, large areas of the country surveyed, many physical observations made, coal found in considerable quantities, and deposits of iron, copper, and gold discovered.

BEKE: *De Veer's Three Voyages of Barents* (Hakluyt Soc. 1876); SPORER: *Nowaja Semla* (Erganz.–Heft. Petermann's Geogr. Mitth. Gotha 1865); MARKHAM, A. H.: *Voyage to Novaya Zemlya*, 1879 (London 1881); LAMONT: *Yachting Voyages in Arctic Seas* (London 1876); NORDENSKIÖLD: *Voyage of the Vega* (New York 1882); TOEPPEN: *Die Doppelinseln Nowaja Semla* (Leipzig, 1878; PASCHOFF (Mme.): *Nossilif's Voyage à la Nouvelle-Zemble* (Tour du Monde, Paris, 1894.)

CHAPTER IV

THE NORTHEAST PASSAGE

THE spirit of trade and commerce animated the original promoters of the search for a Northwest Passage, — a berated spirit, that nevertheless is the basis of the material prosperity which fosters civilization.

Attracted by the great profits of the oriental trade, the enterprising merchants of England, as has been pointed out, sought the northwest way to India and China. When this proved impracticable, they turned for a passage to the coast of the Old World.

The European part of the Northeast Passage, from the very situation, must have been known to Norse and Russian mariners from the earliest times. While the correct form of the Scandinavian peninsula was not mapped till 1539, by OLAUS MAGNUS, yet OTHERE had rounded North Cape in the ninth century, and ISTOMA journeyed from the White Sea to Trondheem in 1496.

As the first extended maritime venture by England in distant seas, the enterprise was viewed as one of no ordinary difficulty. Three ships were built and fitted most substantially, while great care was exercised in selecting crews. Sir HUGH WILLOUGHBY, a soldier of singularly energetic character, was given command, with two captains, RICHARD CHANCELLOR, who had given many proofs of high capacity, and DERFOUTH. The ships — *Bona*

Esperanza, 120 tons and 35 souls; *Edward Bonaventure*, 160 tons, 50 souls; and *Bona Confidentia*, 90 tons, 28 souls — sailed from Radcliff, 30th May 1553. Their departure was marked by salutes and other demonstrations of public approval and enthusiasm, in which the common people and the king's courtiers equally participated. The ships traced in company the coast of Norway around North Cape to Senjen, where CHANCELLOR in the *Bonaventure* was separated from the fleet by a storm. The others proceeded by courses now indeterminate, but it is known that they reached an uninhabited ice-encompassed land, where shallow water made landing impracticable. The shoal sand-banks of Koljugev Island point to this as the shore, although others claim that it wâs Goose Land of Nova Zembla.

Eventually WILLOUGHBY reached, 28th September, the coast of Russian Lapland, and wintered at the mouth of the Varzina, on the barren Arctic coast of the Kola Peninsula. WILLOUGHBY promptly made search in all directions and found the adjacent country uninhabited. The conditions under which they wintered were such that the entire equipage of the two vessels, 62 souls, perished from scurvy. In the ensuing spring the startled Finnish fishermen found the ships in good condition, manned only by a crew of unburied dead.

The more successful voyage of CHANCELLOR proved of vast and untold importance to England. Reaching Vardoe he waited a week for WILLOUGHBY, when, despite the dissuasions of certain of his party, he resolutely determined to proceed, — 'either to bring that to passe which was intended, or else to die the death.' Entering the White Sea, he reached a monastery on the Dwina, where he was received with great hospitality. The Czar, IVAN the Terrible, being informed by courier of his arrival, invited

CHANCELLOR to Moscow. Passing the winter as guest of the Czar, he returned to England in 1554.

The importance of CHANCELLOR's discovery of a route to Russia through the White Sea was obvious to England. Commercial treaties were made which resulted in great and lasting benefits to both nations, and enterprising merchants organized in 1555 the Muscovy Company.

NORDENSKIÖLD says: 'Incalculable was the influence which the voyages of WILLOUGHBY and CHANCELLOR had upon English commerce and on the development of the whole of Russia, and of the north of Norway. From the monastery at the mouth of the Dwina a flourishing commercial town (Archangel) has arisen, and a numerous population has settled on the coast of the Polar Sea. . . . Regular steam communication has commenced along the Arctic Ocean far beyond the sea opened by CHANCELLOR to the world's commerce.'

The company did not relax its strenuous efforts to find a definite route by the northeast to China or India. To this end in the spring of 1556 they sent forth in the *Searchthrift* STEPHEN BURROUGH, one of CHANCELLOR'S companions. On 20th June BURROUGH reached Kola, near which point he separated from CHANCELLOR, who was returning to the White Sea. At Kola were Russian *lodjas*, small rowing and sailing boats, in which these bold mariners ventured northward fishing for walrus and salmon. BURROUGH visited the Petchora, and later, 4th August, anchored between Burrough and St James islands, near the south end of Nova Zembla. To this point he had been accompanied by Russian fishing-boats, without whose assistance and guidance it is doubtful whether he would have reached Nova Zembla. BURROUGH attempted to pass the Waigat Island by the south, but owing to a severe storm, 3d September, returned to

Colomogro for the winter, where he abandoned his contemplated project of reaching the Obi the next summer.

BURROUGH brought to western Europe its first information of Nova Zembla and the Samoyeds (Chapter II). His discovery that the Russian fishing-boats navigated the White Sea, the waters of Nova Zembla, and even at times sailed to the Obi, gave great encouragement to the advocates of the Northeast Passage.

The profits of the Muscovy trade, however, only accentuated the disadvantages under which England and the Netherlands labored respecting trade with the Indies, from which Spain and Portugal were reaping such rich harvests. In the interval between BURROUGH'S voyage to the northeast and the next (PET'S), PHILIP the Second has successfully asserted his claim — in 1580 — to the crown of Portugal, and the entire power of the Iberian Peninsula was exerted against the trade operations of the nations of northern Europe. PHILIP'S arrogance and jealousy reached such a pitch that in 1584 he prohibited the Netherlands from even trading with Portugal.

In this contingency the Muscovy Company made the first long step toward the Northeast Passage by sending out two able and courageous seamen, who sailed from Harwich, 9th June 1580: ARTHUR PET in the *George*, 40 tons, accompanied by CHARLES JACKMAN in the *William*, 20 tons. PET'S crew consisted of nine men and a boy, and JACKMAN'S of five men and a boy, with which these adventurous men safely navigated this part of the Arctic Ocean with their tiny, ill-found barks. PET made Nova Zembla in the vicinity of South Goose Cape, whence he turned south and, kept off shore by ice, missed Burrough's (Waigat) Strait, but coasting Waigat Island entered the mouth of the Petchora. He entered Kara Sea, according to the accepted belief, through Jugor Schar or Pet Strait, but more probably, as NORDENSKIÖLD advances, through Kara Strait.

These were the first vessels from western Europe that were ever forced into the Kara Sea, and the successful navigation of its ice entitles PET and JACKSON to great credit for their determination and judgment. Their voyage practically ended English exploration to the northeast, as the expedition of WOOD and FLAWS in 1666 ended disastrously by shipwreck on the coast of Nova Zembla.

The successful issue of the Spanish war greatly stimulated commerce and navigation in the Netherlands. Despite the destruction of the Spanish Armada, Spain and Portugal yet barred the way to the Indies by the southern route, and the Dutch now took up the northeast search with great energy. In 1594, three ships were sent out, the first under BARENTS, who attempted the search to the north of Nova Zembla. The two other ships, commanded by NAY and TETGALES, fell in with Russian fishermen, near the Petchora, who advised them as to navigation to the eastward. The Dutch landed on Waigat Island, 10th August, where the Samoyed hunters gave them correct and valuable information as to the Kara Sea, its extent and navigability, which they received with disdain. Eventually the Dutch entered Kara Sea through Jugor Schar, and 20th August reached the Kara River, which they supposed to be the Obi.

Encouraged by these reports of NAY, the Dutch sent seven vessels the following year, 1595, NAY being admiral, with TETGALES, BARENTS, and others as captains, and HEEMSKERCK as lieutenant. The farthest point reached was Staten (Mestni) Island, just within the Kara Sea, whence, as NORDENSKIÖLD says, 'we can state with certainty, from the knowledge we now possess of the ice conditions of Kara Sea, that the Dutch . . . had the way open to the

Obi and Yenesei.' The fishermen of the Waigat informed them that Russian trading-boats sailed yearly past the Obi to the Yenesei, — a report which was confirmed by the Samoyeds who inhabited Taimur Land.

The results of the third Dutch expedition under RYP, HEEMSKERCK, and BARENTS have already been described. The return of HEEMSKERCK from Nova Zembla in 1597 ended farther Dutch search in this direction, as in that year HOUTMAN, returning with the first Netherland fleet from the East Indies, proved that the Dutch marine was able to hold her own.

The unsuccessful voyages of the Dutch and English expeditions, in their efforts to discover a navigable Asiatic sea-route between the Atlantic and Pacific, did not deter Russian explorers. These latter mariners bent their energies to determining the outlines and limits of the northern coast of the continent of Asia, and especially to ascertaining how far along the Siberian shores coastwise navigation was practicable; for in this quarter trade — their inspiration — could be expected.

A more important element in eastward exploration than fishing and the coasting-trade was the onward march of hunters for the sable and other valuable fur-animals, which led the Russians into the Lena Valley by 1627. In 1636 ELISHA BUSA, starting with ten other Cossacks to explore the rivers falling into the Polar Sea, reached in successive voyages (1637–1639) the Lena Delta, the Olenek to the west and the Yana to the east. About 1640 POSTNIK discovered overland the river Indigirka; and as early as 1644 the Kolyma Valley, yet farther to the east, saw the erection of a trading-post under the Cossack STADUKIN, destined to be the centre of Russian influence and tribute-exactions. Here STADUKIN acquired such knowledge of the Polar Sea, the New Siberian Islands, and a river

(Anadyr) flowing into the South (Pacific) Ocean as led to extensive explorations. Although details are wanting, it is known that journeys were made to the new Siberian Islands from the Kolyma westward to the Lena, and that SIMEON DESCHNEF, with three other Cossacks, making a sea-voyage from the Kolyma to Kamchatka, in 1648, discovered Bering Strait, and unconsciously determined the non-continuity of the continents of Asia and America.

At this time five-sixths of the Arctic coasts of Europe and Asia had been traversed, all indeed except the shore-line between the Petchora and the Olenek rivers. The knowledge gathered was indeterminate and disconnected, so in the great survey, 1725–1742, the work was done anew in voyages and journeys which will here be very briefly summarized, except that of BERING (Chapter VI).

In 1735–36, MURAVIEF sailed from Archangel to the vicinity of Cape Jamal, which promontory was doubled by his successors, MALYGIN and SKURATOF, who reached the gulf of Obi in 1738, while SELINFONTOF by reindeer sledge followed northward along the gulf to White Island (Beli Ostrov). This practically connected Archangel and the Piciani, 47 degrees of longitude apart, as in the same summer OFTSIN and KASHELEF sailed from the Obi to the Yenesei, and MININ from the Yenesei passed beyond the Piciani, to Cape Sterlegof, 75° 26′ N.

Meanwhile two expeditions sailed from Irkutsk down the Lena, in 1735, one under PRONCHISTSHEF, who reached the vicinity of the northern point (Cape Chelyuskin) of Asia, barely failed of rounding it, and returned to the Olenek. Here terminated mournfully an Arctic romance of peculiar interest. PRONCHISTSHEF took with him on this voyage his newly married bride, who, sharing her husband's perils and sufferings, died within two days of his death on reaching the Olenek. She was then the only recorded white woman to reach such a high latitude, 77° 48′ N., and held that peculiar honor until Madame D'AUNET visited Spitzbergen in 1839.

Lieutenant CHARITON LAPTIEF took up the westward voyage in 1739, under orders from the Admiralty to complete the survey by sea or by land. From his farthest, Cape Thaddeus, 76° 47′ N., he turned back and wintered on the Chatanga. The following year yielded no results save the besetment and loss of his vessel, three hundred miles from his old winter-quarters, their only hope of safety. This journey was made on foot in the beginning of an Arctic winter, across a desolate, uninhabited tundra. Day by day men broke down under excessive exertions and insufficient food, and no less than twelve men perished of cold and exhaustion. Nevertheless, LAPTIEF continued his work by land the following spring, his endeavors being particularly directed to passing around the north point of Asia, — an exceedingly difficult task from its extremely high latitude, in 77° 34′, over six degrees to the north of the extreme northerly point of America, Boothia Felix, 72° N. His mate, CHELYUSKIN, reached this cape, which properly bears his name, in May, 1742, by a long and difficult sledge journey from the Chatanga, while LAPTIEF and others explored the rest of the peninsula and ended voyages to the west. CHELYUSKINS's discovery of the northernmost point of Asia has been often questioned, but NORDENSKIÖLD puts his claims beyond cavil.

The westerly voyage of LASSINIUS from the Lena, in 1735, ended a short distance to the east of the delta in utter disaster, 53 of the 62 men, including LASSINIUS, dying of scurvy in winter quarters. DMITRI LAPTEIF, being then charged with the work, pursued it with such perseverance, resolution, and fidelity, as enabled him, in a succession of dangerous and difficult voyages, 1737–

1742, to skirt the coast through 30 degrees of longitude, from the mouth of the Lena to Cape Baranoff, and reach overland the valley of the Anadyr.

Thus gradually the northern coasts of Asia came within the knowledge of man, while the question of a Northeast Passage passed into the accepted list of impossibilities.

The dormant question, however, was reopened some twenty years since through the exertions of a man whose interests and activities in Arctic matters were such as to insure a scientific and vigorous attempt to solve the problem; this man was ADOLF ERIK NORDENSKIÖLD. Born in Finland, 1832, his Arctic enterprises have covered a greater field than those of any other explorer, and his labors have been productive to an unprecedented degree of results important to science, and beneficial to mankind. From Bering Strait to Greenland, from Spitzbergen to Nova Zembla and Northern Asia, his voyages have invariably promoted the interests of science, advanced commercial projects, and stimulated co-laborers.

When the question of a Northeast Passage was revived in 1875, NORDENSKIÖLD's experiences included the Swedish Arctic expeditions of 1858, 1861, 1864, 1868, 1872, and a voyage to Greenland in 1870, which with other later journeys are mentioned in their appropriate places.

A preliminary journey being advisable, in 1875 Mr OSCAR DICKSON equipped the *Proven*, 70 tons, manned it with twelve walrus-hunters, and put it at the disposal of NORDENSKIÖLD, whose scientific staff included KJELLMAN, LUNDSTRÖM, THÉEL, and STUXBERG. Visiting Nova Zembla and making a rich harvest of scientific information, NORDENSKIÖLD passed through Jugor Strait, 2d August, and despite occasional delay by ice, anchored in the mouth of the Yenesei, at Dickson Harbor, 15th August. The vessel returned to Tromsoe under KJELL-

MAN, while NORDENSKIÖLD, LUNDSTRÖM, and THÉEL ascended the Yenesei and returned overland.

NORDENSKIÖLD'S success in inaugurating a sea-route to Siberia resulted in his receiving the thanks of the Russian Government, but the complete success of this voyage was not convincing to some, who urged that it resulted from an unusual favorable ice season. To meet these objections a second expedition was sent to the Yenesei in 1876, at the expense of Mr DICKSON and Mr ALEXANDER SIBIRIAKOFF; the land expedition was under THÉEL, while NORDENSKIÖLD, in the *Ymer* with KJELLMAN and STUXBERG, left Tromsoe 25th July. The ice conditions of Kara Sea were unfavorable, but NORDENSKIÖLD overcame all obstacles, reached the mouth of the Yenesei 15th August, and returning was in Norway 22d September.

These voyages not only proved the practicability of summer navigation from Western Europe to the Yenesei, but also resulted in extremely valuable scientific collections and observations from the Nova Zembla archipelago and the Siberian coast. The Kara Sea proved rich in individuals and in types, yielding nearly 500 species; from Nova Zembla the species of known insects were raised from 7 to 100, and the knowledge of the vertebrate world of this region was similarly extended.

Thus NORDENSKIÖLD'S explorations on the north coast of Asia have not been barren geographic successes, but have resulted in benefit to Siberia and Western Europe.

Repeated commercial voyages have since been made, — especially a series by Captain JOSEPH WIGGINS, — which have been marked by various measures of success, but which as a whole have proved the Arctic route to be practicable and profitable. In 1889, two ships and four tugs left England at the end of July, and reached Karaul, 160 miles up the Yenesei, in 39 days, and three weeks later made their return voyage in 26 days. This voyage NORDENSKIÖLD describes 'as an event rivalling in importance the return to Portugal of the first fleet loaded with merchandise from India.'

Not content with having inaugurated a sea-route of incalculable value for the development of Northern Asia, NORDENSKIÖLD addressed the Swedish Government a memorial setting forth the practicability of a voyage from the Yenesei to Bering Strait. He set forth the important scientific benefits that must result from such voyage, and clearly indicated its practicability in this age of steam.

His arguments were convincing, and the greater part of the sum of $100,000 necessary for success was furnished by: Mr OSCAR DICKSON, $60,000; King OSCAR, and Mr SIBIRIAKOFF, $11,000 each. The Swedish Diet voted grants for equipping and provisioning the *Vega*, a steam-whaler of 300 tons burthen. Professor NORDENSKIÖLD intrusted the command of the ship to Captain L. PALANDER of the Swedish navy, who with seven other officers and 21 men composed the personnel. The *Vega* left Tromsoe on 21st July 1878, accompanied by three other ships: the collier *Express* with coal for the *Vega*; the *Frazier* with a cargo for the Yenesei River, and the *Lena*, which was intended to proceed to Yakutsk.

The *Vega* found the Kara Straits ice-free, and here the *Express* transferred her coal, during which time the expeditionary force made ethnographic investigations among the Samoyeds. The ice conditions of the Kara Sea proved favorable, and the *Vega* — delaying for dredging, sounding, and other scientific observations — reached Dickson Harbor at the mouth of the Yenesei, 6th August, preceded three days by her sister ship. On 10th August, the *Vega* and *Lena* steamed northeastward along the Asiatic coast, where they found the shore land laid down much too easterly, — in some cases a distance of 80 miles on the *new* chart, although the *original* maps of MININ, who sailed in 1738–39, from the Yenesei to Cape Sterlegof, proved to be entirely accurate. Delayed somewhat by ice and bad weather at the western entrance to Taimir Bay, the *Vega* pushed on at the first possible moment, and on the 19th of August rounded the northernmost point of Asia and the eastern continent, Cape Chelyuskin, 77° 33′ N., 103° 26′ E., and anchored in the bay to the west of it. NORDENSKIÖLD was greatly elated at the passage of this cape, since it virtually ensured the accomplishment of the Northeast Passage.

Sailing eastward, the following day the *Vega* made its most northerly latitude, 77° 45′ N., and although its progress was impeded by bad water and ice, reached the neighborhood of the Lena Delta. The *Lena* here turned southward to the river, while NORDENSKIÖLD, improving the opportunity of an open sea and favorable weather, continued his voyage toward Bering Strait. Favorable conditions continued to the New Siberian Islands, where the shallowness of the water prevented NORDENSKIÖLD from landing on Liakoff, as he originally contemplated. The Bear Islands were sighted 3d September, and Cape Yakan, on the south shore of Long Strait and the nearest point to Wrangell Land, was passed four days later. For the first time progress was completely stopped on 12th September, at the mountainous promontory of Irkaipi, the 'North Cape' of COOK, where, under similar conditions in 1778, he turned back to Bering Strait.

Gradually working its way southward as ice conditions permitted, the *Vega* reached Kolyuchin Bay, immediately to the westward of Cape Serdze Kamen, on a bright, beautiful day, — the hearts of her crew cheered by the knowledge that they had only to round this promontory to pass safely into the Pacific Ocean. Unfortunately, the ice was packed so closely that further progress was impossible, and here, only 120 miles from Bering Strait, they were destined to remain the next ten months. Cold weather set in immediately, and by 1st October the new ice was strong enough to travel over, but, as whalers frequently navigate this sea far into October, NORDENSKIÖLD was unwilling until the end of the month to believe that winter-quarters were inevitable.

The *Vega* wintered off Pitlekai, a Chukche village in 67° 07′ N., 123° E. An ice observatory for magnetic observations was constructed on land, and a regular routine of scientific investigations was laid out for the winter. The most friendly relations were maintained with the natives, and with their aid sledge journeys were made over the Pitlekai peninsula.

With increasing cold the Chukches gradually moved southeastward from the frozen sea at Pitlekai to the great lagoons of the Naisaka, where a plentiful supply of fish ensured subsistence during the winter. Every opportunity was improved to accumulate knowledge as to Chukche customs and language. During the winter NORQUIST acquired a vocabulary of about 1,000 words. Other members of the expedition made ethnographic collections, and kept up magnetic, tidal, astronomical, meteorological, and other physical observations.

On 18th July 1879, the ice broke up and two days later the *Vega*, rounding East Cape with flying colors, saluted the easternmost point of Asia in honor of the completion of the Northeast Passage. Pt Clarence, St Lawrence, and the Commander Islands were visited, and 2d September 1879, the *Vega* anchored at Oklahama, whence NORDENSKIÖLD sent forth to the civilized world the welcome tidings that after three centuries of effort the Northeast Passage had been made in a single voyage.

LESLIE well proclaims NORDENSKIÖLD and his comrades as 'worthy sons of the old Vikings, and as men who had made their names immortal . . . by a splendid victory, achieved by human skill and daring over the powers of nature and the rigors of the Icy seas.'

COXE: *Russian Discoveries* (London 1780); SARYTSCHEW: *Reise im nordöstlichen Sibirien*, 3 v. (Leipzig 1805–15); BURNEY: *North-Eastern Voyages of Discovery* (London 1819); SABINE: *Wrangell's Siberia and Polar Sea* (London 1840); NORDENSKIÖLD: *Voyage of Vega* (New York 1882); *Wissentschaftlichen Ergebnisse der Vega Expedition* (Leipzig, n. d.).

CHAPTER V

SPITZBERGEN

THE archipelago of Spitzbergen is for several reasons the most interesting of Arctic lands. It is the largest known region of the uninhabited earth, it lies intermediate between the Old and New Worlds, its seas have been the richest in exploitable values, its shores enjoy a climate unequalled for its mildness in such latitudes, and it has served as a base for a larger number of Arctic expeditions than any other country. The flora is extensive, and among its fauna, reindeer were once so plentiful that in many successive years the Russian and Norwegian hunters have killed them by thousands annually.

While Spitzbergen is uninhabited, neither the rigors of the climate nor the lack of available subsistence forbid its colonization. Until 1830 the Russians had sent for many years parties to pass a single winter, so as to be on the ground for early spring hunting. SCHAROSTIN passed 15 consecutive years on the island, and spent 17 other winters, till he died at Bell Sound in 1826.

The earliest wintering party, nine English sailors left by mischance, all perished. Four Dutchmen passed the winter safely in 1633, but at the same station four others perished the following year, which ended that attempt at colonization. Other parties have wintered, — some voluntarily, others shipwrecked, — with various degrees of fortune or disaster.

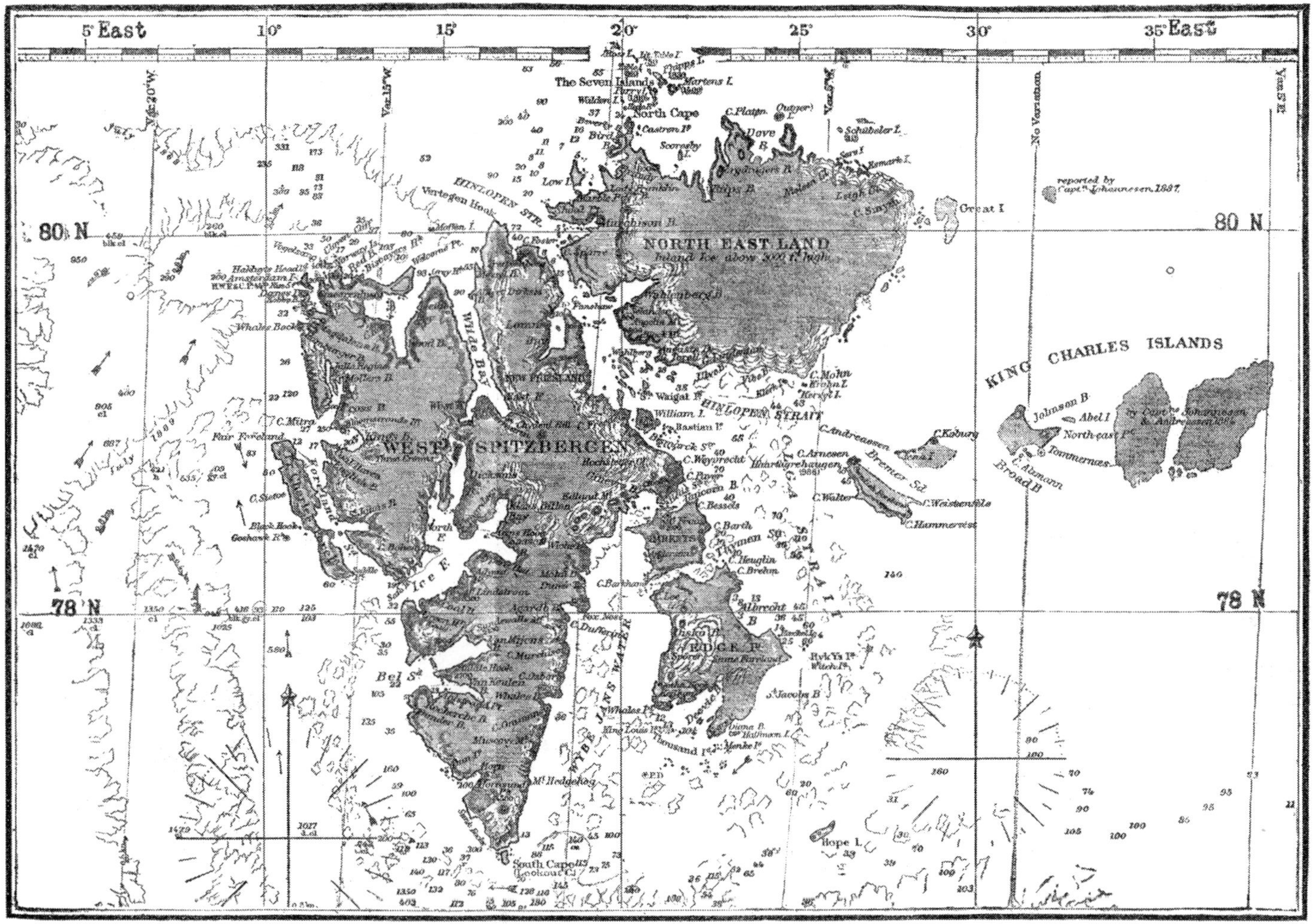

SPITZBERGEN ARCHIPELAGO (HYDROGRAPHIC CHART, U.S.N.)

The first successful wintering is that related by EDWARD PELHAM in "God's Power and Providence; Shewed, in the miraculous Preservation and Deliverance of eight Englishmen, left by mischance in Greenlan (so Spitzbergen was then called) Anno 1630, nine moneths and twelve days. With a true relation of all their miseries, their shifts and hardship they were put to, their food, etc., such as neither Heathen nor Christian men ever before endured." Their courage, ingenuity, and industry brought them safely through, but they were favored by being at Bell Sound, the station of the English whale fleet, where there was a large, well-built house, timber in quantity, whale flesh available, and both sea and land game plentiful; besides, they were well-armed.

The most remarkable experiences are those of four Russian sailors, cast away on the desolate east coast, where they remained seven years, 1742–49. In addition to their usual suit of clothing they had on landing one gun and a few rounds of ammunition, which was speedily used in killing a few deer. They constructed a hut of the plentiful drift-wood that lines that coast, and cast about for means and methods to preserve life. Their experience in this polar land exceeded the romance of Robinson Crusoe. Searching the wood cast on the shore they shaped from suitable material arrows and spears, which were tipped with whalebone, and even a harpoon that was completed from their now useless guns. These were supplemented by bows that were strung with twisted entrails of the slaughtered reindeer. Suitable traps were devised for catching the blue fox, and nets for snaring the waterfowl. They labored not only with energy and industry, but with a definite purpose of acquiring stores of fur and bone of commercial value.

So assiduous and successful were they in the chase and

hunt that yearly there were made large additions to their stock of skins of polar bears, reindeer, seals, and fox. The sixth year almost discouraged them, when one of their number died, but the three others were rescued the following year, when they had saved enough to ensure a considerable sum of money to each man.

The most recent attempt to colonize Spitzbergen occurred in 1872, when Sweden and Norway consulted with other nations with the view of taking possession of the whole country, but Russia objected. The project, based on the reports of Professors NAUCHORST and WILANDER in 1870, looked to the exploitation of phosphatic deposits at Cape Thorsden, where a Swedish company built a house, constructed a small railway, and commenced mining; however, unprofitable results caused abandonment the same summer.

Various claims of no moment have been set up to the discovery of Spitzbergen. The Russians were said to have frequented it for fishing prior to WILLOUGHBY's voyage, and the latter to have reached it in 1553. The former claim seems most improbable, and the latter was clearly set up as a ground for claiming an English monoply in the valuable rights of its adjacent seas, which had been granted the English Muscovy Company in 1613.

The discovery of Spitzbergen is to be ascribed to WILLIAM BARENTS and to JOHN CORNELIUS RYP, who commanded one of the two ships sent by the city of Amsterdam, 10th May 1596, to accomplish the Northeast Passage. The second vessel under HEEMSKERCK had as chief pilot BARENTS, the navigator already celebrated for having reached the north point of Nova Zembla.

After attaining the 71st parallel RYP kept a course northeast by north, which caused BARENTS to remonstrate as being far too easterly for Nova Zembla. The historian of the voyage — DE VEER, who sailed with BARENTS — says: "We being not able with many hard words to purswade them, altered our course . . . to meete them."

On 9th June they discovered an (Bear) island, 74° 30′ N. 18,° 40′ E. Here they tarried four days, finding many birds and a white bear so large that, discomfited they returned from the first attack for reinforcements. Setting on the swimming animal with two boats, they "fought with her two hours, . . . and amongst the rest of the blowes that wee gave her, one of our men stroke into her backe with an axe, which stucke fast in her back, and yet she swomme away with it; at last wee cut her head in sunder with an axe."

Their course hence is a matter of doubt. Peterman and Beke believe that following the east coast they circumnavigated Spitzbergen. This seems most improbable, not only on account of the extreme difficulty of such a voyage, which has never been made in modern times, but also, to my mind, from contemporaneous cartographic data; moreover, they mistook Spitzbergen for a part of Greenland, which would not have happened did they know it to be an island.

The New Map of 1600, part herewith reproduced, (p. 11) shows only the west coast, as does the map of HONDIUS, in PONTANUS' *History of Amsterdam* (1611), constructed to illustrate BARENTS' voyage.

HESSEL GERRITSZ, in his *Histoire du Pays, nommé Spitzbergen* (Amsterdam 1613), publishes BARENTS's log, which runs thus: —

(5th June 1596.) "From Bear Island we set out, shaping our course W. N. W., made 64 miles . . . and N. W., 60 miles. June 4, made N. 1/4 W., 88 miles; . . . we fancied we could see land to the north but were not certain. [This was probably Prince Charles Foreland.] June 15

. . . Sailed until noon S. E., 20 miles, having attained 78 1/4 latitude." This position could not have been on the east coast, nor could the land which was sighted 17th June be made by sailing 16 miles S. S. W. from latitude 80° 10'.[1]

Bear Island was reached 1st July, when RYP and BARENTS, being still unable to agree on their course, parted company. The latter sailed for Nova Zembla (Chapter III), while RYP, returning northward again, successfully visited the north coast of Spitzbergen, without adding to former results.

The discovery of Spitzbergen excited little interest at the time, but it was prominently brought to the attention of the world by the first voyage of HENRY HUDSON, elsewhere described at length (Chapter ——) as North-polar. His northerly progress prevented by an ice-barrier trending to the northeast, HUDSON approached Spitzbergen, which was laid down on his chart, keeping a lookout for Voegel Hook of BARENTS, — presumably the northwestern point of Charles Foreland, — 78° 53' N.

HUDSON made this large island 28 June, 1607, and after circumnavigating it sailed northward and reached the northeastern part of Spitzbergen, *Nieuland or the lana under* 80 *degrees* of BARENTS. Sailing eastward from 13th to 15th July, HUDSON explored its inlets and islands and named several points, one being Hakluyt Headland. Shaping his course to the northeast he reached by observation 80 ° 23' N. latitude, and two days later 81° N., by dead

[1] DE VEER makes the date 19 and the latitude, 80° 11', which corrected for error pointed out by Beke would be 79° 49' N. The land high, snow-covered, and extending some thirty miles nearly E. and W., was probably Red Bay, so that they had rounded West Spitzbergen. Inspecting the islands near Hakluyt headland by boats, they made their way to the W. and S. and visited a number of inlets, probably Magdalena Bay, Ice Fiord, and Bell Sound.

reckoning. He turned back 16th July, on which day HUDSON says, "wee saw more land . . . trending north in our sight, . . . stretching farre into 82 degrees." BEECHY supposes this land to be Seven Islands, in 80° 40′ N., which is undoubtedly HUDSON'S farthest point, for his latitudes were not always accurate in that day of rude instruments. Returning by the west coast he explored a part of Bell Sound, of which he gives an excellent description, Ice Fiord, and other estuaries.

It is interesting to note that HUDSON'S Spitzbergen and other discoveries appeared first on the Hondius map published in 1611, at Amsterdam, to illustrate the third voyage of BARENTS, during which this archipelago was originally discovered.

HUDSON'S voyage was of vast industrial and commercial importance, for his discovery and reports of the vast number of walruses and whales, that frequented the seas, gave rise to the Spitzbergen whale-fishery.

The voyage of POOLE for walruses and exploration, in 1610, was followed by the establishment of the whale-fishery by EDGE in the following year. Enterprising Holland sent its ships in 1613, bringing in its train later whalers from Bremen, France, and other maritime centres.

The whale-fishery, as the most important of Arctic industries, — from which Holland alone drew from the Spitzbergen seas in 110 years, 1679–1778, products valued at about ninety millions of dollars, — merits brief attention.

GRAD writes: 'The Dutch sailors saw in Spitzbergen waters great whales in immense numbers, whose catch would be a source of apparently inexhaustible riches. For two centuries fleets of whalers frequented its seas. The rush to the gold-bearing placers of California and the mines of Australia afford in our day the only examples at all comparable to the host of men attracted by the Northern fishery.'

England viewed with alarm the competition of the Dutch, and, in the interests of the Muscovy Company immediately claimed a monopoly of these fisheries, on the ill-founded assertion that WILLOUGHBY had discovered Spitzbergen in 1555. The English fleet seized by force the first Dutch ships, but worsted in an engagement with a Hollandish whale-fleet in 1618, England agreed to a compromise, under which the Spitzbergen harbors were allotted equitably to various nations. The extreme north fell to Holland and the south to England, with France, the Hanse towns, and others intermediate.

During the most profitable period of the Dutch fishery, 1620–1635, it is within bounds to say that over 300 Dutch ships and more than 15,000 men annually visited Spitzbergen; more than 18,000 men were on the coast in one summer, says LAMONT. It is definitely known that 188 whalers congregated at one anchorage in 1689; and in 1680 the Dutch sent out 260 ships, and about 14,000 men, who made a catch valued at nearly a million and a quarter of dollars.

About 1620, whales frequented the bays and immediate coast of Spitzbergen in such numbers that the fishers were embarrassed to transport homeward the blubber and other products. These conditions led to the summer colonization of Spitzbergen (and Jan Mayen), where establishments for trying-out, cooperage, etc., were erected, as the most economical method of pursuing the industry. They were occupied only in summer, although the experiences of PELHAM and other English sailors, who involuntarily wintered in Spitzbergen 1630–31, led to an attempt to establish a Dutch colony. The party of 1633-34 wintered successfully, but that of the following year perished, and so ended the experiment.

The most remarkable of these establishments was at Amsterdam Island, where on a broad plain grew up the astonishing village of Smeerenberg. Here, nearly within ten degrees of the North Pole, 79° 50′ N., for a score of years, prevailed an amount of comfort and prosperity that can scarcely be credited by the visitor of to-day. Several hundred ships, with more than 10,000 men visited it annually. These consisted, not alone of the whalers and land-laborers, but of the camp-followers who always frequent centres of great and rapid productivity.

In train of the whalers followed merchant-vessels, loaded with wine, brandy, tobacco, and edibles unknown in the plain fare of the hardy fishers. Shops were opened, drinking-booths erected, wooden (and even brick) tile-covered houses constructed for the laborers or visiting whalemen. Even bakeries were constructed, and, as in Holland, the sound of the baker's horn, announcing hot, fresh bread, drew crowds of eager purchasers. If report errs not, even the Dutch frau of 1630 was sufficiently enterprising to visit Smeerenberg, 79° 50′ N., and take away the credit of the farthest north from her Russian sister (p. 41) of 1735 . . . and her French rival, Madame D'AUNET, of 1839 (p. 57), 79° 35′ N.

The shore fisheries soon failed (about 1640) and the Dutch being driven to the remote and open seas, Smeerenberg fell into decadence; the furnaces were demolished, the copper chaldrons removed, and the tools and utensils of the cooper and whaler disappeared; only the polar bear remained to guard the ruins of the famous Spitzbergen fair.

MARTENS says of the few houses remaining in 1671: 'They are built after this fashion: . . . there is a stove before with a ceiling at top, and behind a chamber taking in the whole house: . . . An anville, smith's tongs and other tools frozen up in the ice (were visible) and the kettle was still standing as it was set.'

But human interest in Smeerenberg did not pass away with its vanishing habitations, for on the shores of that bay rest the last mortal remains of a thousand stalwart fishers, who closed their lives of toil and struggle in view of the icy seas that had often witnessed their triumphs over the mighty leviathan of the deep.

Storm-stayed and ice-beset no longer, their dust awaits the change and fate ordained by God's eternal laws. The aspect of this most northerly cemetery of the world finds its parallel in other Spitzbergen harbors. That in Magdalena Bay, 79° 35′ N., is thus described by Madame D'AUNET, who visited it in the *Recherche*, 1839. 'I counted fifty-two graves in this cemetery, which is the most forbidding in the wide world; a cemetery without epitaphs, without monuments, without flowers, without remembrances, without tears, without regrets, without prayers; a cemetery of desolation, where oblivion doubly environs the dead, where is heard no sigh, no voice, no human step; a terrifying solitude, a profound and frigid silence, broken only by the fierce growl of the polar bear or the moaning of the storm.'

The adventurous whalers naturally visited all ice-free regions in favorable years, and in this way the western, northern, and possibly the Hinlopen Strait shores were more or less accurately surveyed. POOL in 1610 named Bell Sound, Ice Fiord, and other points, and BAFFIN, in his second voyage with whalers to Spitzbergen, visited and named Wiches Sound (Wilde Bay), and Sir THOMAS SMITH Inlet (Hinlopen Strait). PHIPPS' north-polar voyage, 1773 (Chapter XII), added nothing of importance to a knowledge of Spitzbergen. His scientific observations and collections were comparatively scanty, and BEECHY characterized his surveys as incorrect and his charts as dangerous for navigators. Indeed, the account of Spitzbergen written by MARTENS, 1671, was not excelled until SCORESBY published his *Polar Regions*, in 1823.

GRAD admirably sums up this account as follows: 'SCORESBY surpasses in extent, variety, and exactness everything written regarding the polar physics up to the beginning of this century. Seventeen voyages to Spitzbergen enabled this gifted observer to fully describe such phenomena as are peculiar to these islands; his book yet remains the initial point of all scientific polar research.'

The Buchan and Franklin expedition (Chapter XVII), 1818, contributed pendulum and other physical observations made at Dane Island, where CLAVERING and SABINE followed them in 1824. PARRY's supporting party at Trurenberg Bay (Chapter XII), 1827, added materially to the scanty stock of physical knowledge of northern Spitzbergen, while in the same year the Dane, KEILHAU, was investigating the geology of the southern coasts.

Next in order France sent to Spitzbergen, in *La Recherche*, Captain FABVRE, a scientific committee, of which GAIMARD, MARMIER, BRAVAIS, MARTINS, and others were members. They occupied Bell Sound in the summer of 1838, and Magdalena Bay, 79° 35′ N., the following year, when the party was accompanied by Madame D'AUNET. A great amount of valuable work, from which all branches of science profited, was done by the commission, but the most interesting was that of CHARLES MARTINS, whose investigations of glaciers formed an epoch in the study of these phenomena. MARTINS's studies led to his brilliant generalizations on present and past floras.

To no country more than to Sweden — with its generous merchants and able scientists — is due the credit of investigating the natural history and the physical conditions of the Spitzbergen archipelago. A series of expeditions covering a quarter of a century was initiated, and in a large part conducted, by OTTO TORELL, chief geologist of Sweden, who, in 1858, sailed on the *Frithiof*, with two assistants, — one of whom then commenced his successful career of Arctic investigation, — A. E. NORDENSKIÖLD and QUENNERSTEDT. Two months were spent in dredging the sea and scouring the land of the west Spitzbergen coast, especial attention being paid to the geological formations, glacial phenomena, and botanical work. NORDENSKIÖLD distinguished himself by his discovery, by a snow-shoe journey, of rich fossil-bearing rock in carboniferous formations, and TORELL's visit to Horn Sound and Amsterdam Island resulted in a remarkable work, *Mollusc-fauna of Spitzbergen, etc.*

TORELL's next expedition, 1861, contemplated a comprehensive survey of the geology and natural history of Spitzbergen archipelago, with geographic explorations to the north and northeast; also determination of the practicability of measuring an arc of meridian in Spitzbergen. The northward voyage was utilized in dredging the sea and in geologizing such land-points as were touched on their journey to Trurenberg Bay, where TORELL landed after an unsuccessful attempt to force a passage around the extreme north of Spitzbergen. Physical observations and research filled in the month of June, during which they were imprisoned by ice in this bay. The enforced delay caused the abandonment of the proposed sledge journey to the far North, owing to unsuitable ice-conditions for sledging. In consequence, TORELL, NORDENSKIÖLD, and PETERSEN undertook a boat voyage through

Hinlopen Strait, while CHYDENIUS made a successful preliminary survey for measuring an arc of meridian.

Later the same party visited the hitherto unexplored coast of Northeast Land, and passing North Cape, 28th July, visited the Seven Islands, and reached their farthest, Phipps Island, 5th August, 80° 42′ N. The 'distant highland' of PARRY (1827) was reached 12th August, and named Prince Oscar Land, and the day following, from a mountain 2,000 feet high, near Cape Wrede, there were visible two islands named Charles XII and the Lifeguard (Drabanten). Their farthest was east of Cape Platen, where ice conditions obliged their return.

Unfortunately, NORDENSKIÖLD'S next visit to Spitzbergen was in 1864, instead of 1863, when the ice conditions were exceedingly favorable for navigation. In this year (1863) the Norwegian CARLSEN circumnavigated Spitzbergen, the first time the feat was accomplished. This voyage was made from the west, which makes the surmise that BARENTS came up the east coast more improbable than ever. Coming from the south the highest latitude on the east coast is that of the Norwegian NILSEN, in the *De Freia*, 79° 20′ N., in 1872.

The Swedish expedition of 1864 was placed under NORDENSKIÖLD'S leadership, and with him were associated DUNÉR and MALMGREN. Their boat was one of only 26 tons burden, with provisions for less than six months, there being neither room nor funds for more supplies.

Forced to enter Safe Harbor at the mouth of Ice Fiord, they explored considerable portions of this magnificent inlet, and later examined portions of Bell and Horn sounds. Unable on account of the pack to go north, they rounded the southern cape of Spitzbergen, and in turn entered Store Fiord, and visited Edges Land and Barents Land. Entering Helis Sound they ascended White Mountain, from which at a long distance to the east was visible very high land. They had discovered a new land or else re-discovered the western part of the land, which, first found by EDGE in 1613, had been completely forgotten by some, its existence denied by others, and which found no place in the new maps. Unable to proceed toward this interesting land, owing to the ice, they again rounded South Cape, intending to follow the west coast as far north as possible. At Charles Foreland, however, they fell in with six boats filled with shipwrecked sailors coming from the north. It appears that three walrus-hunters, TOBIAS, AARSTROM, and MATHILAS, reached the Seven Islands 3d August, when the open sea tempted them to sail around the north point of Northeast Land and down its east coast. Here they were beset, and, as the ice drew them steadily to the south, were obliged to abandon their vessels. By boat they reached NORDENSKIÖLD via Hinlopen Strait, travelling 200 miles in 14 days. Farther explorations were thus precluded, but NORDENSKIÖLD had the great satisfaction arising from his instrumentality in thus saving human life. Following this expedition, NORDENSKIÖLD and DUNÉR produced a map — based on 80 different points, exactly determined in this and other expeditions by geodetic methods — which delineates Spitzbergen with an accuracy unattained as regards any other Arctic land.

The Swedish north-polar expeditions of 1868 and 1872 are elsewhere described (Chapter XII), but the latter had a more intimate relation with Spitzbergen than utilizing it as a base. NORDENSKIÖLD occupied Mussel Bay for his winter station, where his two provision convoys, *Gladen* and *Onkle Adam*, were ice-bound by a violent storm, 16th September 1872. No provision for wintering these vessels had been made, and NORDENSKIÖLD and PALANDER, his captain and navigator, found themselves with 67 mouths to feed, instead of 24, which could only be done by reducing the rations of all by one-third.

This course had hardly been decided on and winter quarters begun, when they were startled and dismayed by the arrival of four men with news that six walrus-vessels had been frozen in at Point Grey and Cape Welcome, west of Wilde Bay. There were 58 of these unfortunate men, and their provisions could be eked out by hunting, it was thought, till 1st December; after that starvation impended, — unless NORDENSKIÖLD helped them. To attempt to feed 125 men on provisions that were insufficient for 67 could only involve all fatally, but to refuse help was impossible to a man like NORDENSKIÖLD.

Two additional sources of food were available. That year a Swedish colony was initiated at Cape Thorsden, Ice Fiord, where a railway and houses were built for working phosphatic deposits. Owing to circumstances, the manager had been obliged to abandon work and return to Norway, leaving behind a considerable amount of provisions. To reach these stores entailed a journey of nearly 200 miles, but the attempt must be made by some of the walrus-hunters. The hunters agreed to make this effort. The second possibility was to involve for many months daily sacrifices on the part of the Swedes in their attempt to utilize a large stock of reindeer moss as an article of food, which was done by baking moss and rye-flour mixed, thus producing a very bitter bread, which was, however, eatable. The moss was brought for the draught-animals, reindeer, of the north-polar sledge party, and its conversion into food meant the sacrifice of his cherished plan of exploration, to which NORDENSKIÖLD had given several years of his life.

It may here be said that 17 walrus-hunters, under the veteran MATHILAS, reached by boat Cape Thorsden, where they found all the necessaries of life, — house, fuel, working tools, preserved and dried vegetables, and fresh potatoes. They lived, however, principally on salt-beef and pork, remained inactive and packed themselves in one small room. This neglect of hygienic laws was followed by scurvy, and every man died.

The situation at Mussel Bay steadily grew worse, for through improvidence or misfortune the remaining hunters informed NORDENSKIÖLD that their food would fail two weeks earlier than had been expected, and then they must depend entirely on NORDENSKIÖLD'S supplies, which were later depleted by the escape, through neglect of the Lapps, of all his reindeer during a violent snow-storm. It was arranged that the Norwegians should join NORDENSKIÖLD 10th November; most fortunately, however, the ice opened so that two of the vessels escaped, taking all the crews of the four others except two men who remained — and died that winter — in charge of the valuable cargoes.

NORDENSKIÖLD neglected nothing that was conducive to the health or morale of the party. Games were played, schools opened, scientific observations made, dredging under the ice taken up, hunting followed (which secured five wild reindeer), and excursions made, which kept the men busy and in a measure healthful. Insufficient food, however, was followed by repeated outbreaks of scurvy, which proved amenable under hospital diet.

When the end of April made travel possible NORDENSKIÖLD attempted his northern journey, but was stopped at Seven Islands by unfavorable ice (Chapter XII). Determined to utilize his opportunities to the utmost, he decided to return round North East Land, in order to determine its geographic structure, to note the position of adjacent islands, to settle the disputed eastern boundary, and to ascertain the character of its inland ice. Five days were taken to pass over the 23 miles intervening between Phipps Island and Cape Platen, the way being over loose angular blocks of ice, piled pyramid-like in heaps ten to thirty feet high. Reaching, 31st May, the extreme eastern end of the mainland, Otter Island offered from its summit, 330 feet high, a view of miles to the east and north where no land was visible. This visit, however, confirmed the accuracy of LEIGH SMITH'S survey by which North East Land was extended considerably to the eastward.

The presence of a large water-hole to the south caused NORDENSKIÖLD to take the inland ice at that point, instead of following the coast to a more southerly point. This ice-sheet, some 2,000 to 3,000 feet thick, flows toward the east and so presents on that coast a precipitous ice-wall, insurmountable from the sea. Unbroken by mountain ridges, its sea-front presents the broadest known glacier.

On 1st June they ascended the inland ice, where their first experiences were full of danger. NORDENSKIÖLD says: 'Scarcely had we advanced 2,000 feet farther before one of our men disappeared, at a place where the ice was quite level, and so instantaneously that he could not even give a cry for help. When we, affrighted, looked into the hole made where he disappeared, we found him hanging on the drag-line, to which he was fastened with reindeer harness, over a deep abyss. . . . He was hoisted up unhurt. . . . If his arms had slipped out of the reindeer harness, a single belt, he would have been lost.'

Of the ice conditions he says: 'Along the level ice-surface every puff of wind drove a stream of fine snow-dust, which, from the ease with which it penetrated everywhere, was as troublesome to us as the fine sand of the desert to the travellers in Sahara. By means of this fine snow-dust, steadily driven forward by the wind, the upper part of the glacier — which did not consist of ice, as in Greenland, but of hard-packed, blinding white snow — was glazed and polished so that we might have thought ourselves to be advancing over an unsurpassably faultless and spotless floor of white marble.'

The journey occupied fifteen days, which were marked by blizzards or ice-fogs, the latter so dense that every hollow had to be tested by lowering a man to determine the depth, 'it being impossible to distinguish by the eye whether we had before us a deep impassable channel or only a depression of a couple of feet.'

Canals were fallen in with, from 30 to 100 feet wide, with parallel vertical walls sometimes 40 feet high. Over these canals were often found snow-bridges, that facilitated travel at the expense of safety; for on one occasion, just as the party were about to cross over a depth that would have proved fatal in case of accident, the bridge fell.

The original plan contemplated crossing to Cape Mohn, but a rugged terrain compelled NORDENSKIÖLD to turn to the west and descend into Hinlopen Strait, at Wahlenberg Bay, which was reached 15th June, when it presented to the ice-weary travellers the first sign of summer in its flowering red saxifrages.

The return journey was extended to nearly twice the time fixed, but the difficulties and dangers met with were more than compensated by the scientific results, especially in regard to the knowledge of the peculiar inland ice so materially different from that of Greenland.

The return of NORDENSKIÖLD, 23d May, was followed before the end of the month by flowering plants, returning birds, and the breaking up of the main ice-pack; and on 12th June the arrival and generous liberality of LEIGH SMITH placed the expedition beyond fear of the future.

The scientific researches connected with the International Polar Station at Cape Thorsden in 1882 (Chapter XVI), and the botanical investigations the same year by NAATHORST and DE GEER and the exploration of the glaciers between Horn Sound and Recherche Bay by NORDENSKIÖLD, Jr. and BJORLING, in 1890, substantially close the record of Sweden's efforts. As to the results, it may be said that they have been not alone of material benefit, but have been morally and intellectually profitable. Fishing interests have been fostered, as for example the profitable catching of the great northern shark, — one of the ten fishes added by MALMGREM to the meagre number of four previously known in Spitzbergen waters. The archipelago has been traversed, examined, and charted in a most satisfactory manner, its geology studied, its plant life determined (the number of known flowering plants being raised from 57 to 96), and its glacial features studied.

Nearly 7,000 printed pages of scientific memoirs and books have appeared, many showing that the general questions of the world need such data for their solution. While the contributions of MALMGREM, NORDENSKIÖLD, TORELL, and others have been valuable, yet an especial interest attaches to the investigations of Professor O. HEER, the author of *Fossila Arctica Flora*. HEER's astute and valuable researches have clearly established the fact that at one time Spitzbergen was covered with a luxuriant miocene vegetation, — cypresses, birches, sequoiæ, oaks, and planes. It moreover appears that this growth was coincident with the period when Spitzbergen, Greenland, Franz Josef Land, and Nova Zembla formed an immense unbroken continent, which inevitably experienced a continental climate.

The visits of LAMONT to Spitzbergen, 1858, 1859, 1869, and 1871, were for the purposes of hunting, but they afforded much information of interest in regard to the animal life of the archipelago. The voyages of LEIGH SMITH, 1871 and 1872, have been of great importance geographically, here as well as in Franz Joseph Land (Chapter XV). In 1871 with Captain ULVE in the *Sampson*, SMITH, tracing the whole northern coast of Spitzbergen and North East Land, rounded the latter and reached Cape Smith on the unknown eastern coast. Later he attained 18° E., 81° 24′ (or 81° 30′) N., one of the highest latitudes ever reached by vessel, to the northwest of Seven Islands, — a close approach to Giles Land. In 1872, as related, he succored the unfortunate Swedes at Mussel Bay.

East of Spitzbergen are two lands, which have caused endless dispute as to their location and nomenclature,— Wiche Land and Giles Land. In 1707 GILES discovered to the east of Spitzbergen, in about 80° N., a land, to which his name was given. We are also told that THOMAS EDGE, commanding the *Muscovy* in 1617, 'discovered to the eastward of Spitzbergen, as far to the north as 79°, an island which he named Wiche's Island.'

In time geographers disputed the existence of both lands, — for to-day we know there are two lands, in 80° and 78°. Wiche Land passed out of mind until seen in 1864 by NORDENSKIÖLD and others of the Swedish expedition from Mount White, and by NEWTON and BIRBECK the same year from the Ryk Isles. Again observed from the high land of Thymens Strait by VON HEUGLIN in 1870, he called it King Charles Land, which seems to be its accepted name. In 1872, three Norwegian whalers — ALTMAN, in July, NEILS JOHANNESEN in August, and NILSEN later — circumnavigated it and landed in 78° 10′ N. It is a mountainous land with scanty vegetation, but enough to support animal life, as shown by the deer there killed by JOHANNESEN. King Charles Land was charted from these reports some eight degrees too far to the east, as shown by observations made at the visit of JOHANNESEN, HEMMING, and ANDREASSEN in 1884, when they also found that instead of one island there were three, thus confirming ALTMAN'S opinion against that of JOHANNESEN. Dr W. KUKENTHAL, while making zoological observations in the *Cecilie Malene*, Captain M. ARNESEN, visited these islands four times in 1889, and explored them fully.

Separated from other lands by depths far greater than the height of its famous Beerenberg mount, isolated Jan Mayen Island — fog-enshrouded and ice-beset — pertains rather geographically to Greenland than to Spitzbergen, with which its history naturally connects it.

The discovery of this island by HUDSON, in 1607, is thus set forth in EDGE'S *Briefe Discoverie:* 'In ranging homewards he discovered an island lying in seventy-one degrees, which he named Hudson's Tutches.' As early as 1618 it appears from state papers that the English whalers frequented it, for they complain that 'Hudson's Touches and adjoining islands' (Egg and Rocky Islets) had been claimed exclusively by the Hollanders. The Dutch held fast both to name and island, where a miniature Smeerenberg grew up, after a fatal attempt in 1634 to establish a winter colony on Jan Mayen. With the decadence of whale-fishery, Jan Mayen was left to its desolation, largely owing to its great inaccessibility, from fogs and dangerous ice. The most important expedition that has landed on the island was in connection with the Austrian-Hungarian international polar station, Captain WOHLGEMUTH, 1882–83.

In 1891 M. CH. RABOT, in *Château-Renauel* landed on the unfrequented east coast, and explored Esk and Faskrud fiords, but in 1892 failed to effect a landing in the *Manche.*

Probably Jan Mayen is best known to English readers by the charming account given by Lord DUFFERIN, in

Letters from High Latitudes of his plucky visit in 1856, when an exploring steamer, *La Reine Hortense*, failed to reach the island. DUFFERIN's description of his first sight of the famous Mt. Beerenberg is worthy of reproduction:

'The heavy wreaths of vapor seemed to be imperceptibly separating, the solid roof of gray suddenly split asunder, and I beheld through the gap, — thousands of feet overhead, as if suspended in the crystal sky, — a cone of illuminated snow. The roof of mist closed again. At last . . . the brown changed to gray, the gray to white, and white to transparent blue at the horizon; but an impenetrable veil hung suspended, behind which I knew must lie Jan Mayen. A few minutes more, and slowly, silently, in a manner you could take no count of, its dusky hem first deepened to a violet tinge, then, gradually lifting displayed a long line of coast, — in reality but the roots of Beerenberg, — dyed of the darkest purple; while, obedient to a common impulse, the clouds that wrapped its summit gently disengaged themselves, and left the mountain standing in all the magnificence of his 5,836 feet, girdled by a single zone of pearly vapor, from underneath whose floating folds seven enormous glaciers rolled down into the sea.'

WHITE: *Spitzbergen and Greenland* (Hak. Soc., London 1855); PETERMANN: *Spitzbergen u. d. Arktische Central Region* (Erganz.-Heft No. 16, Peterm. Mitth. Gotha 1862); GRAD: *Iles Spitzbergen* (Paris 1866); LESLIE: *Voyages of Nordenskiöld*, 1858–1878 (London 1879). For Jan Mayen, see WOHLGEMUTH, Chapter XVIII.

CHAPTER VI

BERING STRAIT

THE much discussed question as to how America was first peopled has for a battle-ground the Bering Strait region. Whether America was peopled from Asia, or the reverse migration occurred, there is no doubt that this narrow channel has for centuries been passed and repassed by the littoral tribes of Asia and America. It is well known that from the beginning of the 16th century the natives of the two continents have maintained intercommunication for barter.

The populations of the two shores are widely separated, from the racial standpoint, the warlike, property-holding Chuckches to the west, the peaceful and semi-communistic Eskimo to the east; both again differing from the Aleut to the south.

There are two classes of Asiatic natives — there being about 2,000 of each — the reindeer and the coast Chuckches. The latter being without deer, live a half-nomadic, half-permanent life along the coast of Bering Strait, where they have intermixed somewhat with the Aleuts and Eskimo. The other division, reindeer-owning nomads, who live by trade and reindeer-raising, wander over the region between the Indigirka and the strait.

The Chukches seem to be the only Siberian race that has had sufficient moral force and physical bravery to withstand the tribute-exactions of Russian officials and

No. V.

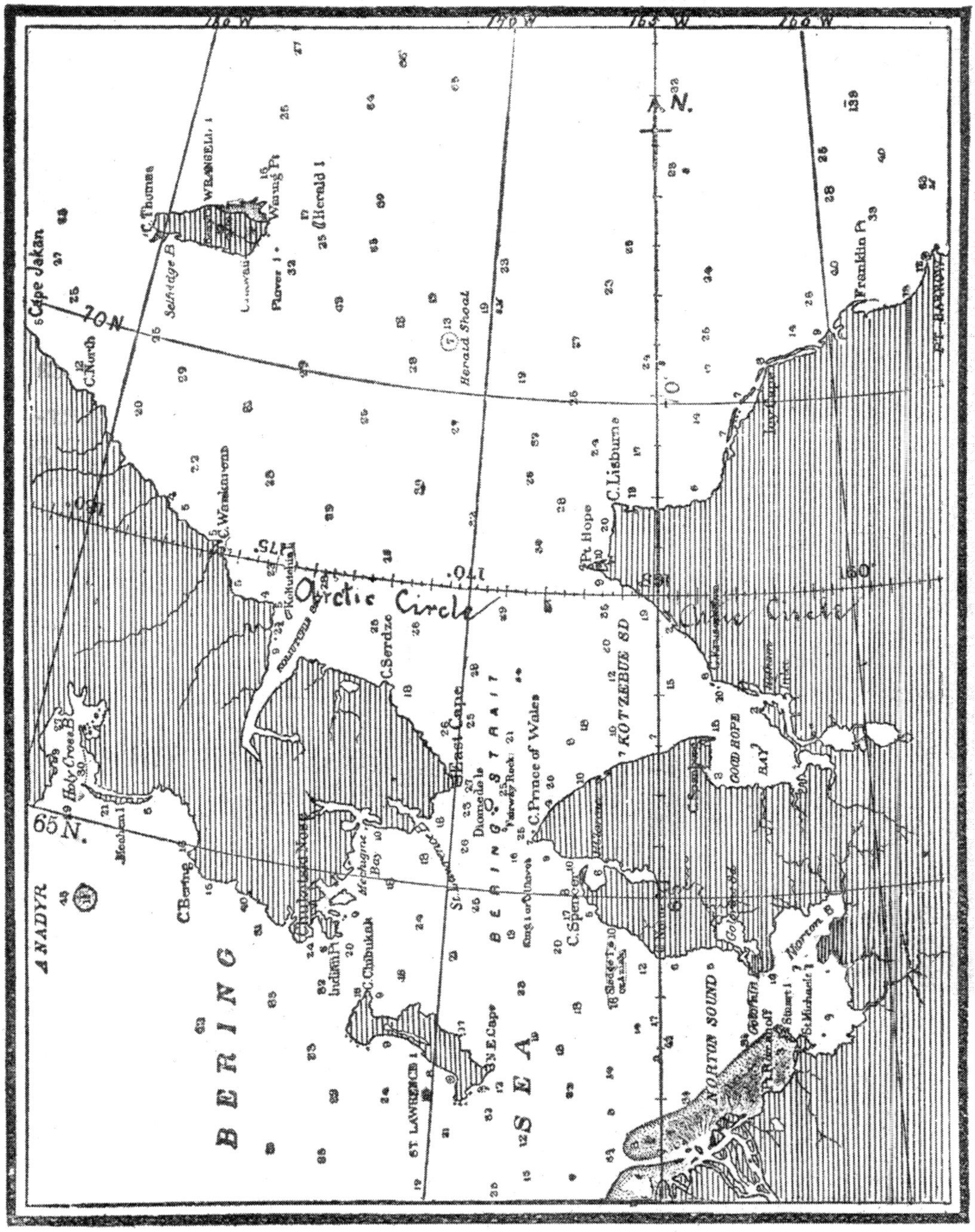

BERING STRAIT AND WRANGELL ISLAND

(Hydrographic Chart, U.S. Navy)

the enslaving methods of the fur-trader. The efforts of the Russians — under SCHESTAKOF, 1730, and PALUTSKI, 1731 — to subdue and enforce tribute from the Chuckches proved costly and fruitless. The adoption of a conciliatory policy and the withdrawal of the enormously expensive garrison of Anadyrsk, resulted in the establishment of peaceful relations with these tribes, which have continued. SARYSCHEF, 1798, WRANGEL, 1823, LÜTKE, 1826–27, HOOPER, 1849, DITMAR, 1853, and others have described the Chuckches before frequent communication with Europeans had modified their environment.

One of the islands of Bering Strait, St. Lawrence, discovered by BERING, 1728, was first visited by BILLINGS, 1st August 1791. The generation seen by KOTZEBUE, 27th June 1816, and 20th July 1817, had never seen Europeans, but they received him with great, indeed embarrassing kindness. Occasionally they are visited by explorers, notably NORDENSKIÖLD, 1880, and in later years by whalers and traders, not always to the advantage of the natives. Walrus frequent in large numbers the island, thus affording food for the improvident natives. In 1880–81 about 200 natives died of starvation owing to illicit whiskey-trading, which diverted them from the hunt when game was plenty.

The first important contribution to a knowledge of the Eskimo of Bering Strait was made by Surgeon JOHN SIMPSON, H. M. S. *Plover*, from his observations 1851–53. He confirms the report of traffic by means of Chukche boats, between (East) Cape Deshnef, Diomede Islands, and Cape Prince of Wales. While the coast near Icy Cape was occupied in 1851 by Eskimo, yet Point Barrow was the favorite camping-ground, from the unusual facilities it affords the entire year for seal-fishing. The establishment of the international polar station at Point Barrow,

1882–83, resulted in very complete and interesting accounts of these Eskimo by JOHN MURDOCH (Chapter XVI).

The action of the United States in building at Point Barrow a relief station for shipwrecked mariners, and the frequenting of the station by whalers and trading vessels have naturally caused material changes in the habits and surroundings of these Eskimo. The condition of the natives near Cape Prince of Wales has been materially improved by the United States, through the efforts and recommendations of Dr SHELDON JACKSON, whereby schools have been opened, and, of more immediate benefit, reindeer imported from the Asiatic coast.

The first passage of Bering Strait was the outcome of commercial enterprises, resulting from the efforts of the Russian traders at the mouth of the Colima to extend eastward their trading operations. The successful voyage of ISAI IGNATIEF in 1646, whereby he visited the Chuckches to the eastward of the Colima and traded for walrus ivory, inspired SIMON DESHNEF to an unsuccessful voyage in search of the reputed Anadyr River.

Renewed efforts in 1648 led to the departure from the Colima of seven vessels. Four were soon disabled, but three, commanded by DESHNEF, ALEXIEF, and ANKUDINOF, found the sea very open, and, rounding successfully the northeast extremity of Asia, passed into Bering Strait. Shipwreck, hostile encounters, and disease destroyed the entire expedition except DESHNEF and his crew, who escaped from the wreck of their vessel on the coast of Kamchatka, south of Anadyr Bay. Supporting themselves by the chase during the winter, on the banks of the Anadyr, DESHNEF's party ascended that stream the following year, and built the post of Anadyrsk, thus founding the first civilized settlement in the Bering Strait region. Traditionally, in 1654, a trader named STADUKIN following DESHNEF'S route, had circumnavigated Cape Deshnef and reached Kamchatka and the Kurile Islands.

Gradually reinforced overland in the coming years by traders, — mostly Cossacks, — the Russians extended their sway over the adjacent regions. In 1711, POPOF, visiting East Cape, or Cape Deshnef as NORDENSKIÖLD proposes to call it, to exact tribute from the Chuckches, brought back an account of islands (Diomedes) and a continent (America) to the eastward, a report of the contiguity of America which was the cause of BERING'S later voyages.

The voyages of VITUS BERING, from whom the strait derives its name, resulted from the action of PETER the Great, who, in the last year of his eventful life, planned one of the greatest geographic expeditions ever recorded. It is known as the Great Northern Expedition, and its execution, after PETER'S death, entailed 17 years (1725–1742) of effort on the part of the explorers, and is said to have impoverished many tribes by its heavy exactions of supplies and unceasing demands for transportation.

The series of voyages and journeys that explored the northern coasts of Asia have been elsewhere recorded (Chapter IV), but they were expeditions supplementary to the first voyage to Kamchatka, which, organized under Empress ANNE, left St. Petersburg 4th February 1725, under command of BERING, a Dane in the Russian service.

The overland journey across Siberia, the accumulation of supplies at Okhotsk and the construction of a vessel at that remote port, involved almost insuperable and innumerable delays. It was 30th June 1727 before SPANBERG, BERING'S assistant, sailed in his newly completed vessel, *Fortuna*, with supplies, material, and workmen for ship-building, across the sea of Kamchatka to Bolsheretsk. Another winter passed in accumulating supplies, getting out ship-timber, and in other essential work at Lower Kamchatka, 56° 15′ N., 162° W. Finally, in June the *Gabriel* was launched, and on 24th July 1728 BERING sailed out of the Kamchatka River, after three years and four months of preparation for a voyage that lasted seven weeks. Coasting northward, he met in about 64° 30′ N., a party of Chukches, navigating the Gulf of Anadyr, who informed him that a short distance to the northeast the mainland took a decided turn, changing its direction to the north and west.

Pursuing his way, the day of Saint Lawrence brought BERING to an island, which he named for the patron saint. Passing East Cape, 26th August, the *Gabriel* coasted along the Asiatic shore to the neighborhood of 67° 18′ N., 170° W. From this point BERING turned back; as he says, 'because the land no longer extended north. Neither from the Chukche coast *nor to the eastward could any extension of the land be observed.* If we should continue on our course and happen to have contrary winds we could not get back to Kamchatka before the close of navigation.' CHAPLIN, his lieutenant, says that BERING turned back 'in spite of his instructions.' In any event, this voyage added nothing to the discoveries of DESHNEF in 1648, since the words above quoted indicate clearly that the American continent to the eastward was not seen, contrary to the opinion sometimes advanced.

Wintering on the Kamchatka River, BERING feebly attempted farther explorations in 1729, sailing 16th June to the eastward in search of land that was said to be visible on fine days. His voyage did not exceed a hundred miles and lasted less than three days, when the search was abandoned. WAXEL says that BERING had no intention of returning to Kamchatka River; however that may be, he did not do so, but instead rounded the south cape of

Kamchatka, and landing at Bolsheretsk 13th July, made his way overland as speedily as possible to St. Petersburg, which he reached 12th March 1730.

While BERING was returning to report the result of his expedition, TSCHIRIKOF was endeavoring to reduce the hostile Chukches; and in a co-operating voyage the navigator MICHAEL GWOSDEF visited, in 1731, Cape Szerde Kamen, whence an easterly storm drove him to anticipate BERING in discovering the coast of America, along which GWOSDEF coasted two days before returning to Kamchatka.

BERING had been absent five years from St. Petersburg, and the essential object of his voyage — to determine whether or not Asia and America were separated — was undetermined, much to the disappointment of its promoters. Despite criticism, the Empress ANNE listened to the representations of BERING, and gave orders for a second expedition on a more extended scale under his command. In the spring of 1733 BERING left St. Petersburg with his former lieutenant, SPANBERG, and other officers. Assigned to the specific duty of making discoveries to the southward, SPANBERG sailed from Kamchatka in 1739, and extended his explorations along the Kurile islands to Japan.

It was not until September 1740, seven years after his second departure from St. Petersburg, that BERING with TSCHIRIKOF left Ohkotsk, and rounding Kamchatka anchored at the port of Avatcha, now Petropaulovsk, where he wintered. On 15th June 1741, BERING's final expedition sailed from Petropaulovsk. With the commander in the *St Peter* were Lieutenant WAXEL and WILLIAM STELLER, the latter a German physician and naturalist of great promise, while TSCHIRIKOF, in the *St Paul*, was accompanied by DE LA CROYERE, an astronomer. They were to sail in company, with the purpose of discovering the northwest coast of America, and of definitely determining whether or not its shores united with those of Asia. Valuable time and favorable weather were wasted in searching to the southeast for the mythical land of Gama. Eventually they turned their course to the northeast, but the eighth day out they parted during foggy weather.

TSCHIRIKOF, sailing to the northeast, reached, 26th July, the American coast in the vicinity of Cross Bay. A violent sea beating against a precipitous, rocky coast, caused him to keep some distance off shore. Unable to make a safe harbor, on the third day he sent his mate, ABRAHAM DEMENTIEF, with ten men to examine the adjacent country. Day after day passed with no sign of the land party, and TSCHIRIKOF, much alarmed at its prolonged absence, sent, 4th August, the boatswain, SIDOR SAFELEF, with three men in his only remaining boat. No man of either party was ever seen again, and their fate is entirely conjectural. Meanwhile continual smoke indicated the presence of natives, who showed themselves, two men in different canoes, 26th July, and after hailing the vessel in an unknown tongue disappeared. The following day TSCHIRIKOF, despairing of again seeing his men, determined to explore as much of the new land as was possible and then return to Kamchatka. He followed the coast more than a hundred miles, hindered by contrary winds and frequent fogs. Once his vessel barely escaped stranding, and he was approached by 21 natives in kayaks, or leather boats holding a single man. They spoke an unknown language, but DE LA CROYERE assured TSCHIRIKOF that they were like the natives he had intimately known in Canada. Lack of fresh water and suitable food was accompanied by outbreaks of scurvy, from which both of his lieutenants and others of his crew died. Under these conditions return to Petropaulovsk was imperative, and TSCHIRIKOF safely entered that port, 21st October 1741, losing by death his astronomer the same day.

BERING'S explorations were much of the same order as those of his comrade. The American coast was reached 29th July, between Capes St. Elias and St. Hermogenes. Here he remained three days. A party was landed for fresh water, an opportunity eagerly improved by STELLER to gain information as to the country and inhabitants. He found freshly cooked food, and sent his Cossack back to bring up a party to follow the traces, but BERING imperatively ordered his instant return to the ship. While STELLER naturally complained of BERING'S timidity, it was fully justified by the fate of TSCHIRIKOF'S parties.

Consulting with his officers, BERING decided to return, and weighing anchor before sunrise, 31st July, followed the coast to the westward. Entangled in the coast-wise islands he took to the open sea, where contrary winds and storms retarded his progress. Water failing, he made an island, named Shumagin, from a sailor he buried there, where the supply taken from a lake proved to be brackish and unhealthy. Frequent camp-fires seen at night caused them to keep a close lookout for natives. None were seen until 15th September, when some kayakers appeared and made friendly overtures toward trade, but misunderstandings soon broke off all commerce. Proceeding two days later, islands were so numerous that BERING then turned more to the south; here, in 51° N. latitude, near a number of coast islands, a violent storm of 17 days' duration left them in sad condition, short of provisions and with only a third of their original crew for duty. Two islands were seen 10th November, when BERING stood to the north, and 15th November found himself in 56° N. The day following the vessel stranded on a sandshoal of a desert (Bering) island, and speedily broke up.

Fortunately the whole party reached shore, and yet more fortunately the remains of their ship were driven up within their reach, which circumstance proved their ultimate salvation, the gathered timbers serving as material for a boat the following spring. There was not a tree on the island, but the drift-wood was so abundant that they built huts.

BERING, previously sick, was now utterly discouraged, and refusing to eat, drink, or accept the shelter of a hut, died 19th December 1741. WAXEL says: 'We younger persons recovered our spirits, took courage, resolved to do our utmost, and leave no means untried to save our lives.' The energy, resolution, and skill of this young officer, on whom the command now devolved, proved equal to the situation and secured the safety of the party. In this he was ably seconded by the professional skill and personal efforts of Dr STELLER, whose industry, cheerfulness, and application were unbounded. He thus sums up the misery of the whole winter: 'Want, nakedness, frost, rain, illness, impatience, and despair were our daily companions.' The ambition of the naturalist did not desert STELLER, and his observations on Bering Island, made under desperate conditions, are most valuable contributions, bearing as they do on the extinct sea-cow and on other species of sea animals previous to their being so closely pursued by hunters. STELLER's labors were repaid by most cruel treatment from the Russian Government. Persecuted and hounded by jealous and suspicious officials till he was broken in health and spirit, he perished miserably in the wilds of Siberia at the early age of 37. It is interesting to note that an American expedition, in 1882–83, visited the Commander (Bering and adjacent) Islands, for research as to their physical conditions, fauna, and flora. The results have appeared in a series of memoirs by LEONARD STEJNEGER admirably treating of these subjects.

BERING discovered neither the sea nor strait that bear his name, his voyage to America was preceded by that of GWOSDEF, and he left the question of the separation of the two great continents as he found it, practically unsettled. Yet his voyages were fruitful in geographic results, for by their accounts the hitherto legendary knowledge of this region was first put in accessible and fairly accurate shape. After saying: 'BERING delineated the (Asiatic) coast very well, and fixed the latitude and longitude of the points better than could be expected from the methods he had to go by,' COOK, the navigator, shrewdly and justly adds: 'His misfortunes proved to be the source of much private advantage to individuals, and of public utility to the Russian nation.'

The wintering at Bering Island proved to be the beginning of a pursuit that has added scores of millions of dollars to the wealth of the world. The island was infested by the rare blue fox in such numbers as to be most annoying to the shipwrecked crew, while the extremely valuable sea-otter were so plentiful and tame that they were killed with the greatest ease. Appreciating the value of these skins, the shipwrecked mariners secured more than 900, and brought with them to the mainland furs to the value of $100,000, — which we may believe, as WAXEL says, 'made us some amends for our sufferings.'

Three years after the return of BERING'S shipwrecked crew, in 1745, MICHAEL NOVIDISKOF, sailing eastward from Kamchatka in an open frail craft, reached Attoo, the most westerly of the Aleutian Isles, in search of the sea-otter. His success excited the cupidity and stimulated the activity of other adventurous hunters and traders.

Possessed by a love of gain that no danger could appall, other Russian traders pushed undauntedly eastward from Kamchatka in their frail shallops, most of them called sewn-boats, moss-calked and merely timbers lashed to frames, until one after one the long line of Aleutian Islands and portions of the Alaskan mainland fell within their knowledge, and experienced their atrocious visitations. Prominent actors in these successes were TRAPESNIKOF, GLOTTOF, and PUSHKAREF, — the last-named wintered on the American continent in 1761, — whose daring in navigation was equalled only by their rapacity in trade and their cruelty toward the harmless natives. Such were the energy and skill of these traders that in 30 years the Aleutian archipelago was explored, and the coast lines of the fringing isles of the continent traced southward to meet the farthest northing of the Spaniard PEREZ at Nootka Sound.

All this exploration drifted to the south. Toward the Arctic Circle the landing of SYND at Cape Prince of Wales, in 1767, appears to have been the only addition until the coming of Captain JAMES COOK. The voyage, in the *Resolution*, of this distinguished navigator called him from the enjoyment of honors flowing from his remarkably successful voyages in the South and Antarctic seas, to attempt a passage from the Pacific to the Atlantic, either to the east or to the west. COOK'S instructions required him to follow the northwest coast to 65° N., without delay, and beyond that latitude to carefully examine all bays or inlets that appeared to promise passage either to Hudson or to Baffin bay. This failing, he was to similarly attempt the Northeast Passage.

Passing and naming Mt. St. Elias in May 1778, COOK entered Bering Strait in August, Captain CLERKE accompanying him in the *Discovery*. The unparalleled successes of COOK in the Antarctic ice and over the broad expanse of the Pacific now failed him in a degree. Despite his most strenuous efforts he was unable to force his way through the ice-pack that crowded the Asiatic coast at Cape Szerde Kamen, which had been rounded by DESHNEF in 1648. On the American side his fortunes were better, for he succeeded in reaching, 18th August 1778, 70° 44′ N., where a compact wall of ice turned him back. He charted Capes Lisburne and Icy, but the fog prevented his seeing the (Kotzebue) sound farther to the south. His farthest was about five degrees of latitude to the north of SYND. Returning southward, Norton Sound was explored and named, and the mythical island of 'Alaschka' charted by STAEHLIN, became the mainland of America. COOK intended to renew the attempt in 1779, but his unfortunate death at Hawaii, 14th February 1779, left CLERKE in command, who, death-struck by consumption, died at Petropaulovsk after failing to reach the 70th parallel at his farthest (69° 30′ N.), 22d July 1779. This ended exploration in this region for half a century.

In 1816 a Russian, OTTO VON KOTZEBUE, in the *Rurik*, very appropriately passed through Bering Strait, and made important discoveries. Tracing the American coast northward from Cape Prince of Wales, KOTZEBUE entered, 4th August, the great sound that now bears his name. Delayed here two weeks by his careful survey of the inlet, KOTZEBUE made no higher northing than Cape Krusenstern, 67° N., but his geological discoveries were of especial interest and importance. In Escholtz Bay, 66° 16′ N., he found an ice-cliff 80 feet high, covered by a thin layer of blue clay and turf-earth. From the ice-cliff were obtained bones and tusks of the mammoth, under conditions that lead to the belief that the ice-strata were formed during the life of the mammoth.

The co-operating expedition of BEECHEY (Chapter VII), 1826, for tracing with FRANKLIN the northern coast of America, succeeded in passing somewhat farther to the north than COOK, reaching at sea 71° 08′ N., 163° 40′ W., 13th August 1826. BEECHEY surveyed Kotzebue Sound, while Mate ELSON in a barge succeeded in following the coast 126 miles beyond Icy Cape to Point Barrow, 71° 24′ N., 156° 22′ W., — which was long thought to be the most northerly point of the continent of America. As is now known, this honor lies with the north point of Boothia Felix (of Captain JOHN ROSS), in 72° N.

BEECHEY seems to have been the first to question the influence of Bering Strait as a material feeder or discharger of the Arctic Ocean, for he notes that no 'great body of water flows toward Bering Strait' from the sea of Kamchatka. He adds: 'It appears that near the strait, with southerly and easterly winds, there is a current to the northward, and with northerly and northwesterly winds there is none to the southward; consequently the preponderance is in favor of the former.'

While the voyage of LÜTKE, in *Le Semiavine*, 1828–29, did not result in additional discoveries, yet he charted with accuracy a considerable part of the coast of Asia, between latitudes 53° and 65° N.

The search expeditions for the missing vessels of Sir JOHN FRANKLIN — *Herald*, Captain KELLETT; *Plover*, Captain MOORE, and afterwards Captain MAGUIRE; *Enterprise*, Captain COLLINSON; and *Investigator*, Captain McCLURE — made only incidental contributions to a knowledge of the Bering Strait region. The voyage of Lieutenant JOHN RODGERS, in the *Vincennes*, 1855, resulted in a series of soundings through the strait, supplemented by astronomical, ethnographic, and other observations by Lieutenant BROOKE at Glassecap, 65° N., 172° 35′ W., on the Asiatic coast. The success of RODGERS in attaining the highest latitude made to that time in the Bering Strait region, is set forth in Chapter XIII.

In 1865–66, expeditions visited Bering Strait in connection with the project of establishing telegraphic communication between America and Asia by a cable across the strait. A careful hydrographic survey of the strait, between 64° and 66° N., showed that it is very shallow, with an equable depth of about 20 fathoms.

While the various expeditions already named contributed more or less information bearing on the hydrography of Bering Straits, it remained for WILLIAM H. DALL — the well-known authority on parts of Alaska, without the scope of this volume — to make an exhaustive survey of the strait, and determine a number of mooted questions. During 1871–74, the U. S. Coast Survey work by DALL in Bering Sea comprised many thousand temperature and current observations, and the establishment of a series of magnetic stations from Sitka west to the Aleutian chain, by which the fact was developed that the secular change of the magnetic declination in this region had not only reached its eastern elongation, but had materially receded from the values determined by RODGERS in 1855. In 1880, in the *Yukon*, DALL, supplementing these observations by a series carried northward, by stations at suitable points in Bering Sea and Strait and northward to Point Belcher, fully confirmed the conclusions previously reached. A hydrothermal section of Bering Strait was made, with serial temperatures every 5 fathoms at intervals of 4 miles, from Cape Prince of Wales to East Cape. This work showed the highest temperature to be 48° near the American coast, gradually cooling to 36° near the Asiatic side. The current observations (taken in connection with those previously made north and south of the strait) showed that the principal current in Bering Strait is tidal and intermittent; that the warmth of the water is due to the warming up by the sun of the shallow waters of Norton and Kotzebue sounds, and not to a warm ocean current from the southern part of Bering Sea; that the water south of St. Lawrence Island in Bering Sea is constantly colder than that on the eastern side of the strait and in the two sounds; and finally, that the chief current of Bering Sea and Strait is a feeble but somewhat general movement of cold Arctic water southward. The hypothetical branch of the Japan current, which had long been supposed to enter Bering Sea and extend through the strait into the Arctic Ocean, was conclusively shown to have no actual existence, and the fact that the current in the strait runs northward with the flood, and southeastward with the ebb tide was demonstrated by the observations.

The contributions of the international polar stations maintained at Point Barrow, 1881–83, are set forth in Chapter XVI. They comprise the discovery of Meade River, ethnographic studies, and other scientific investigations, and practically close the exploration of this region.

KOTZEBUE: *Voyage of Discovery*, 1815–18, 3 v. (London 1821); BEECHEY: *Voyage to Bering Strait*, 2 v. (London 1831); SEEMANN: *Voyage of the Herald*, 2 v. (London 1853); HOOPER: *Tents of the Tuski* (London 1853); LAURIDSEN: *Life of Vitus Bering* (Chicago 1890); SIMPSON, MAGUIRE, and MOORE: see *Arctic Blue Books*, Chapter XVIII.

No. VI.

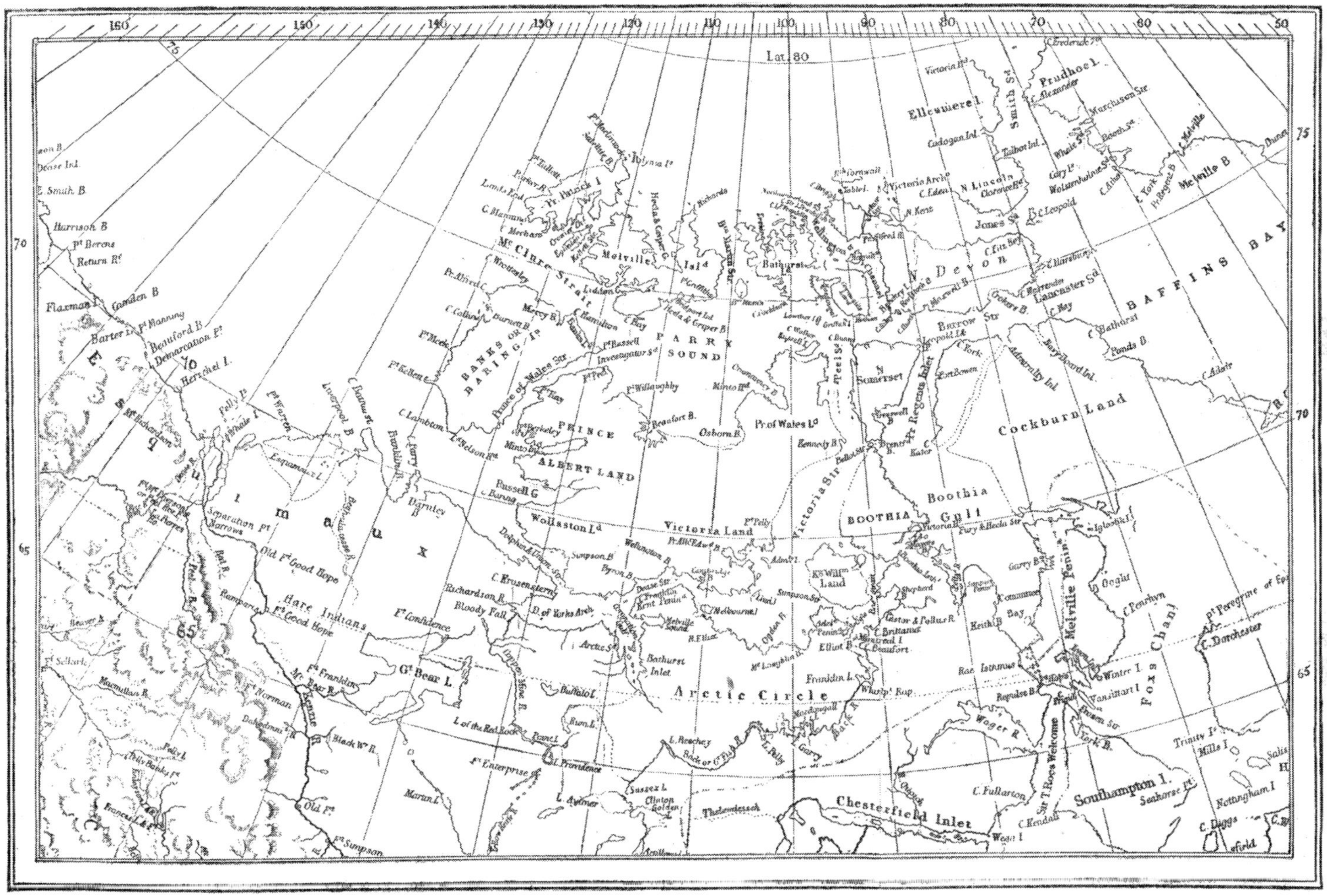

ARCTIC COAST AND ISLANDS OF NORTH AMERICA (AFTER ARROWSMITH)

CHAPTER VII

THE NORTHWEST PASSAGE BY SEA.

AS related already, the efforts to find a Northwest Passage to Cathay and the Indies practically ended with the discoveries of BAFFIN in 1616. The last voyage of BERING and his successors, about the middle of the 18th century (Chapter VI), demonstrating that America and Asia were separated by a strait, excited anew the interest of Great Britain in the forgotten passage. The great navigator, JAMES COOK, called into service for the sole purpose of making this circumnavigation of the Americas, attacked the problem from the new quarter, via Bering Strait. COOK only succeeded in reaching Icy Cape to the east, whence returning for the winter to Hawaii, the spear of a savage ended his notable career.

American and European wars interrupted for 40 years all British exploration, when its renewal was due to the representations of a Scotch whaler, WILLIAM SCORESBY the younger, whose discoveries on the east coast of Greenland are mentioned in Chapter XVII. In 1817, when transmitting to Sir JOSEPH BANKS, some results of his scientific discoveries in the Arctic seas, SCORESBY set forth that he had found 'about 2,000 square leagues of the Greenland (or Spitzbergen) Sea, between 74°, and 80° north, perfectly void of ice . . . whereby (he) was enabled to penetrate within sight of the eastern coast of Greenland, to a meridian usually considered inaccessible.'

He then recommended 'the examination of the coasts of Spitzbergen and Greenland, explorations affecting the whale-fishery, and researches toward deciding whether or not a navigation into the Pacific by a northeast or northwest passage existed,' and tendered his services for such explorations. Through the exertions of Sir JOSEPH BANKS and Sir JOHN BARROW, then Secretary of the Admiralty, the British government decided to renew polar exploration, and in 1818 sent out two expeditions, one to the northwest, and the second to reach the pole. Attached to these expeditions were six men whose names and fame must ever be associated with Arctic discoveries, — BACK, BEECHEY, FRANKLIN, PARRY, JOHN ROSS, and JAMES C. ROSS.

The Northwest Passage was sought by Captain JOHN ROSS in the *Isabella*, seconded by Lieutenant W. E. PARRY in the *Alexander*. As the land expeditions of HEARNE and MACKENZIE had determined that there was no passage below the Arctic Circle, Ross was directed to proceed up Davis Strait to a high latitude, and, stretching to the westward, thus attempt to reach Bering Strait.

The then accepted geographic knowledge placed an imaginary island (James) in the middle of Davis Strait, while the northerly extension of water into Baffin Bay was not credited. BARROW, in his *Arctic Voyages* (London 1818), gives a brief account of BAFFIN'S remarkable voyage of 1616, but does not credit the story, saying, 'It is most vague, indefinite, and unsatisfactory, and the account is most unlike the writing of WILLIAM BAFFIN,' and then proceeds to erase Baffin Bay from his chart, as did EGEDE in his *Greenland*, 1818. BARRINGTON, the same year, in *Possibility of Approaching the North Pole Asserted*, enters the bay with the legend 'According to the relation of W. BAFFIN, 1616, but not now believed.'

But while the expert map-makers were terminating the west coast of Greenland with the 73d parallel, hardy and enterprising whalers, British and American, were yearly visiting its dangerous waters, where they had explored the coast to near the 76th degree of north latitude, and ventured their ships into the ice-clad sea, to the neighborhood of Cape York, within an equal distance of the farthest of Baffin. In 1817 Captain WM. BRASS, whaler *Thomas*, reached 75° 10′ N.; Captain MUIRHEAD, of the *Larkin*, attained 72° 12′ in the open sea; and doubtless other whalers had gone yet beyond.

JOHN ROSS sailed from Lerwick, 3d May 1818, and overtook the ice-bound whaling fleet at Hare Island, 70° 43′ N., 57° W., where he found no less than 45 sail. Following their advice to hug the Greenland shore, Ross pushed northward, aided and accompanied by the whalers, the *Bon Accord* leaving him only in 75° 32′ N., 60° 22′ W., within about 75 miles of Cape York.

Near Cape York both the *Isabella* and *Alexander* were beset by heavy ice during a gale, whereby they were badly injured and barely escaped destruction, their safety depending largely on their special strengthening for ice-navigation. West of Bushnan Island Ross fell in with natives, the Cape York or Etah Eskimo, to whom he gave the fanciful name of Arctic Highlanders. They were 18 in number, provided with dogs, sledges, hunting-gear and iron knives. These last were made of meteoric iron, of which large masses were said to be situated at Sowallick, near Bushnan Island. Rounding Cape York, Ross discovered the red snow (*Protococcus nivalis*) of Cape Dudley Digges, the glacier of Petowick, and reached his farthest point, 76° 54′ N., to the northwest of Carey Islands. Here he saw to the northeast Hakluyt Island of BAFFIN. Falling in with ice, he erroneously concluded from a survey 'that the land is here continuous, and there is no opening at the northernmost part of Baffin Bay.' Turning southwestward he failed to penetrate Jones Sound, but farther to the south reached Lancaster Sound. This strait he penetrated some 50 miles, where he made the astonishing error of thinking this waterway, 30 miles wide, was only a bay surrounded by mountains. Abandoning his only chance of making the Northwest Passage, Ross returned to England, much to the chagrin of PARRY. The voyage had not been fruitless. Ross had confirmed the discoveries of BAFFIN, determined the non-existence of James Island, extended Lancaster Sound, charted the west coast of Davis Strait, made a valuable series of magnetic, meteorological, and hydrographic observations, and added a new folk to the knowledge of the world. His failure in explorations, the main work, was so marked, however, that a five-years Arctic voyage in later years hardly rehabilitated his standing among Arctic men.

The dissatisfaction with Ross's non-success was so pronounced in Great Britain that a new expedition was determined on within a month after his return, and its command given to his lieutenant, WILLIAM EDWARD PARRY, who proved to be one of the ablest, as he was one of the most successful, of Arctic explorers. In command of the *Hecla*, 375 tons, with his assistant, Lieutenant LIDDON, in the *Griper*, 180 tons, PARRY left Yarmouth 12th May 1819, with orders to find the Northwest Passage. Associated with him for scientific work was Captain SABINE, the most persistent and successful magnetic observer of his day. Reaching 73° N., on the west coast of Greenland, PARRY boldly decided to force a passage direct to Lancaster Sound through the ice that covers Baffin Bay, now known as the Middle Ice. PARRY, entering Lancaster Sound on 1st August, a month earlier than Ross, fortunately found open water. The mirage mountains of the previous year had vanished, and as PARRY crowded sail westward, he opened a series of magnificent waterways hitherto unknown. The way lay through an archipelago (Parry) with North Devon, Cornwallis, Bathurst, and Melville islands to the north, and Cockburn, Prince of Wales, and Banks islands to the south. Lancaster Sound, broken at its western end by Prince Regent Inlet, gave way to Barrow Strait, which broadened into Melville Sound, while yet farther to the west the encroaching land formed Banks Strait, wherethrough these channels open into the Polar Ocean.

In his westward voyage PARRY'S scientific assistant was indefatigable in magnetic observations, which assumed an importance before unknown. The route nearly followed the 74th parallel of latitude, and as they went west the declination (variation from pointing due north) of the magnetic needle, which was 109° W. at the mouth of Lancaster Sound, steadily increased to 129° W., at Cornwallis Island. At Byam Martin Island, 75° N., 104° W., SABINE'S observation showed that the declination was 166° east, while the freely suspended compass-needle stood practically upright, its inclination or dip being within 1.6° of vertical. The expedition had thus passed north of the magnetic pole, from east to west, and at some point travelled over, probably Bathurst Island, the compass needle must have pointed due south instead of due north.

On 4th September 1819, PARRY, who later reached 113° 48′ W., passed the 110 meridian west of Greenwich, and thus earned the reward of £5,000, authorized by act of Parliament for such success. Stopped by heavy ice, but confident of making the passage the next year, PARRY made his winter quarters on Melville Island, 74° 47′ N., 114° W.

A land journey in June enabled PARRY to determine the

limited extension of Melville Island to the north. In late June the harbor ice broke up, and on 8th August the impatient explorers sailed westward to attain Bering Strait, but they only reached 114° w. long. High land (Banks Land) existed to the southwest, and the intervening passage (Banks Strait) into the Polar Ocean was blocked up with ice far exceeding in thickness any they had before seen. The bergs, doubtless offshoots from the ice-clad regions north of Parry Archipelago, were from 40 to 100 feet or more thick, and of a hardness commensurate with their size. Frequent efforts to penetrate the pack barely failing of destroying the *Griper*, PARRY reluctantly decided, his officers all advising, to return to England. The homeward voyage was marked by no matter of interest except the discovery of the Eskimo settlement of Possession Bay, at the mouth of Lancaster Sound.

PARRY reached England in November 1820, where he was received with great enthusiasm, as he well might be after a voyage of unprecedented Arctic success since the days of DAVIS and BAFFIN, unmarked by fatality, — except the death of one man from a non-Arctic cause. He had carried the English flag more than half way from Greenland to Bering Strait, passed to the north of the continent of America and of the north magnetic pole, discovered a great archipelago, and encouraged the extension to Lancaster Sound of the whaling industry, worth hundreds of pounds sterling yearly to England.

Uncertain as to the success of his land co-laborer, — FRANKLIN, — PARRY concluded that the ice conditions between Melville Sound and the unknown coast of North America must be even more favorable than by the route he had so advantageously followed. The Admiralty fitted out another expedition at once, and PARRY sailed in May

1821, in the *Fury*, with Captain GEORGE F. LYON, in the *Hecla*. The officers, carefully selected, were men of marked ability, — such as JAMES CLARK ROSS, HOPPNER, NIAS, REID, BUSHNAN, and FISHER, of previous Arctic service, and CROZIER, who later perished in the Franklin expedition. The *Fury* and *Hecla* were twin ships, with interchangeable fittings, one of the many ingenious ideas for Arctic success that PARRY evolved in his five polar voyages. The official instructions directed PARRY to penetrate to the westward through Hudson Strait, reach the coast of the continent of America, and following it northward seek a passage to the westward from the Atlantic to the Pacific.

After leaving Hudson Strait PARRY crossed Fox Channel, and proceeding to the north of Northumberland Island, reached Repulse Bay through Frozen Strait. Since the time of MIDDLETON, 1742, it had been asserted and controverted that the way of the Northwest Passage was through this supposed strait. PARRY settled this controversy of a century's standing by a survey that showed it to be completely landlocked. Two weeks were lost in this side expedition, for it was not until 23d August that the squadron emerged from Frozen Strait and resumed their voyage northward through Fox Channel. Six weeks more were lost through bad ice and the examination of the coast from Repulse Bay to Hoppner and Lyons inlets, the latter promising at first a western passage. On 2d September, after the *Fury* had been nearly shipwrecked, lost an anchor, and been driven south, PARRY says: 'After laborious investigations which have occupied one month, we have by unavoidable circumstances returned to nearly the same spot we had been on August 6.' The rapidly lengthening nights, new ice, and increasing cold forced PARRY to seek, 8th October 1821, winter quarters, at Winter Island, Melville Peninsula, in 66° 32′ N., 84° W.

Of all things that engaged their attention the visits of adjacent bands of Eskimo were fullest of resource and interest. Visits were interchanged, mutual assistance rendered, and cordial relations maintained. Seal and walrus hunting afforded subsistence, food, and shelter to the natives, whose methods of hunting are described by PARRY.

Nor was the knowledge of these people confined alone to hunting, for under questioning and urging they drew maps of the adjacent country that proved surprisingly accurate. The undiscovered strait of Fury and Hecla to the north, Committee Bay to the west, and Rae Peninsula to the south of Melville Peninsula were correctly charted, although the proportionate distances were somewhat inaccurate. A few sledge journeys, made under difficulties enhanced by severe gales, only served to confirm the information given by the Eskimo.

With a grateful sense of relief PARRY saw the land ice break up, 12th July 1822, and permit him to sail northward. His discovery of Hecla and Fury Strait confirmed the Eskimo report as to its existence, but an unbroken floe in its western portion left to conjecture whether or not it led to the western polar sea. Parry entertained no doubt that such was the case; we now know that he was mistaken, as it debouches into Committee Bay, whence Boothia Felix Peninsula extends far northward.

A second wintering at Igloolik, 69° 22′ N., at the east entrance of the strait of Hecla and Fury, brought his crew out in such enfeebled condition that, when the ice held firm in the strait, he was obliged to forego his intention of sending the *Hecla* back, while he remained for another year with a picked crew in the *Fury*. Reluctantly accepting the advice of his surgeon, the two ships reached Lerwick, 10th October 1823.

The failure of PARRY's second voyage impaired neither the confidence of the great navigator nor that of the Admiralty, in the possibility of discovering the long-sought passage, and 19th May 1824 was marked by his departure in command of a new expedition, the *Hecla* and *Fury*. Old associates and experienced officers were with PARRY, — HOPPNER, AUSTIN, JAMES CLARK ROSS, FOSTER, HOOPER, CROZIER, and RICHARDS.

The passage was now to be attempted by the southern inlet (Prince Regent) which PARRY had discovered in 1819, when its ice-encumbered waters forbade exploration. He confidently hoped that favorable weather would remove this barrier during the navigable season, and relying on his judgment the Admiralty authorized the attempt. This opinion was justified by PARRY's experience in his first voyage, when he sailed 120 miles southward in the inlet, to 71° 53′ N., whence, he says, 'I saw no reason to doubt the practicability of ships penetrating much farther to the south by watching the occasional openings in the ice. It is also probable that a channel exists between the western land (North Somerset) and the northern coast of America.' The ice conditions of Baffin Bay proved so unfavorable that, despite desperate attempts, PARRY did not reach Lancaster Sound until 10th September. The season had so far advanced that the new ice most seriously impaired their progress, although the sound was free of old ice. The new ice forced PARRY into winter quarters in Prince Regent Inlet, at Port Bowen 73° 12′ N., very greatly to his disappointment. He had failed to reach by 80 miles his farthest southing of 1819, and as to Port Bowen he had said: 'This spot was the most barren I ever saw.'

Their winter's experience confirmed this opinion as to the desolation and barrenness of the region, and with the greatest delight they quit the harbor on 12th July 1825, to pursue their southward voyage along the west coast of Prince Regent Inlet. Heavy ice and severe gales, however, forced the *Fury* ashore four times, entailing such injuries that after 25 days of effort to save the ship she was abandoned, her stores being left on the ship or beach, where later they proved invaluable to JOHN ROSS; and her crew returned in the *Hecla* to England, where they arrived 12th October 1825.

In this journey PARRY formulated the well-known canons regarding ice-navigation, which time and experience have only tended to confirm. He says: 'The eastern coast of any portion of land, or, what is the same thing, the western sides of seas or inlets having a tendency at all approaching north and south, are, at a given season of the year, generally more encumbered with ice than the shores with an opposite aspect.' Ships, he adds, should be kept disengaged from ice so that they may be 'at liberty to take advantage of the occasional openings in-shore, by which alone the navigation of these seas is to be performed with any degree of certainty.'

This voyage ended PARRY's efforts to make the Northwest Passage, but in 1827 he attempted to reach the North Pole, with results given elsewhere (Chapter XII).

In 1824, the British government decided to supplement the discoveries of PARRY's third expedition, by sending a ship to winter at Repulse Bay, whence the following spring a sledge party should cross the narrow (Rae) isthmus, which the Eskimo said led to a western sea, and follow the coast to Point Turnagain of FRANKLIN. The *Griper* was assigned to the duty, under Captain G. F. LYON, an officer who had navigated these waters with PARRY. Instead of following the northern and short route to Repulse Bay, by way of Frozen Strait, which PARRY had selected in 1821, LYON followed the longer southern way through Hudson Bay to Rowe's Welcome. The selection proved disastrous, for the Welcome was so filled with heavy ice that the *Griper* could not force her way northward. Two violent storms were here experienced, from both of which the vessel escaped shipwreck almost by miracle. In the latter storm, 13th September, all her bowers parted and the *Griper* was thrown over on her broadside, but the wind changing she escaped with the loss of all her anchors. Under these conditions LYON returned to England, his farthest being off Wager River, some 80 miles south of Repulse Bay.

The fruitless efforts of these repeated northwest voyages so discouraged the Admiralty that no farther efforts were made by them until the voyage of BACK, in 1836. Private enterprise in this direction was, however, not wanting, and in 1829 another expedition sailed in the paddle-steamer *Victory*, fitted out by the liberality of FELIX BOOTH, sheriff of London, and commanded by Captain JOHN ROSS. Since his unfortunate mistake of 1818, ROSS had striven in vain to obtain command of another government expedition. An officer of great gallantry and acknowledged ability, yet ROSS'S Arctic fitness was questioned, and his effort viewed with prejudice, owing largely, it would seem, to want of tact, which led to endless controversies. However, both HOPPNER, PARRY'S lieutenant, and BACK, sought service in vain with ROSS on this voyage, wherein he was seconded by his nephew, JAMES CLARK ROSS. The *Victory* was a small paddle-steamer of 150 tons, but this first application of steam power to Arctic exploration did not result favorably, owing to the crudity of the mechanism, and the useless and despised engine was eventually thrown away. Sailing from Woolwich, 23d May 1829, the *Victory* crossed Baffin Bay without difficulty, and on 10th August had passed through Lancaster Sound to the mouth of Prince Regent Inlet. Following its western land, Ross fortunately found it comparatively ice-free, and landing at the beach where the *Fury* had been wrecked, found that the vessel had disappeared.

From the stores which PARRY had evidently landed, Ross filled the *Victory* with canned meats, vegetables, powder, and all kinds of supplies, which later proved his salvation. Passing Cape Garry, the ultimate discovery of PARRY, Ross traced North Somerset to its southern extremity. He was, however, ignorant of this fact, and with phenomenal lack of perspicacity, followed the shore of Creswell Bay, and crossed a (Bellot) strait without recognizing it as such. Calling the indentation Brentford Bay, he passed unconsciously by the northernmost point of the continent of America and the strait that was the Northwest Passage, the object of that and his previous voyage. This passage remained unknown a quarter of a century longer, until one of Ross's successors, pursuing the work of exploration perhaps with a keener instinct and larger capacity, and certainly with more efficient means and facilities, spied it out. The sharp sailor eye of KENNEDY, searching for FRANKLIN in the *Prince Albert*, 1851–52, discovering it, he named it for BELLOT, the French volunteer who lost his life in 1853 as INGLEFIELD'S assistant.

Ross appropriately gave the new land the name of Boothia Felix, after his patron, and tracing southward its desolate shores, finally was driven to (Felix) harbor, 1st October 1829, in 69° 59′ N., 92° W.

The Eskimo in the neighborhood of their winter-quarters proved friendly and serviceable. Learning from them that the western sea was separated from their own water only by a narrow isthmus, 40 miles to the southwest, Ross sent his nephew, JAMES C. ROSS, into the field for land explorations. From a series of journeys it developed that Boothia Felix was a part of the mainland of America, and was connected with it by an isthmus about 15 miles wide.

Young Ross's explorations were extensive and important. Crossing the isthmus he discovered Franklin Passage, Victoria Strait, and King William Land. He followed up the west shore of Boothia Felix Land, to Cape Nicholas the First, about 70° 30′ N., and to the westward, rounding the northern cape of King William Land, traced its western coast to Point Franklin, within 222 geographical miles of FRANKLIN's Point Turnagain. Failing to note Rae Strait, he did not recognize the insularity of this land. The most important work done by JAMES CLARK ROSS, giving imperishable renown to his name, was the determination of the position of the north magnetic pole, which his observations placed at Cape Adelaide, on the west coast of Boothia Felix, in latitude 70° 05′ N., longitude 96° 44′ W.

Unfavorable ice conditions made it impossible for Ross to extricate the *Victory*, and after three winters, 1829–32, failing food made it evident that the lives of the party rested on a retreat to Fury Beach, where the stores cached by PARRY would ensure their safety. Abandoning their ship 29th May 1832, after a voyage of some 300 miles, which bad weather and unfavorable ice-conditions greatly prolonged they reached Fury Beach, 1st July, in dire extremities of hunger and fatigue. Recuperating they attempted to reach Lancaster Sound, hoping to fall in with whalers, but the impenetrable ice to the northward drove them back to winter quarters at Fury Beach, in a house of their construction, called North Somerset. The ice conditions of 1833 were more favorable, and in a voyage of several weeks they crossed Navy Board Inlet, and 25th

August fell in with the whaler *Isabella*, the ship in which Ross had made his first voyage.

The experiences, duration, and results of this voyage are among the most extraordinary on record. The party passed five years in the Arctic regions without fatality, save three (two from non-Arctic causes), discovered a new land, the northern extremity of the continent of America, and made other extensive geographical discoveries. Its observations are probably the most valuable single set ever made within the Arctic Circle, involving not only the climatic conditions of Arctic America, a local matter, but also the determination of the magnetic elements at their very poles, a subject of world-wide importance.

The unpopularity of Captain JOHN ROSS resulted in attempted detractions from the true merit of his work, which was credited in an undue degree to his nephew, JAMES C. ROSS. Fame and reputation enough for any two men should accrue from such accomplishment as resulted from the energy, application, and ability displayed on this memorable expedition.

In 1836, the Royal Geographical Society petitioned the British government to send an expedition to survey the coast between Regent Inlet and Point Turnagain of Franklin, 1821. The *Terror* was placed under command of Sir GEORGE BACK, an able officer of great Arctic experience, who had orders to proceed to Wager River or Repulse Bay, and having crossed Regent Inlet, examine the coast-line east to Cape Kater and west to Back River.

BACK sailed, 14th June 1836, and found the ice conditions of Hudson Bay bad beyond description. The *Terror* was beset the middle of September, near Cape Bylot, and thenceforth was subject for ten months to the vicissitudes of the moving pack. Again and again they passed through most trying ice-experiences, during which instant destruction threatened the vessels for hours at a time. With reference to one occasion, when the ice, thrown up in all directions, overhung the ship, BACK says: 'Though I had seen vast bodies of ice from Spitzbergen to 150 west longitude, all more or less awe-inspiring, I had never witnessed, nor even imagined, anything so fearfully magnificent as the moving towers that frowned on every side.'

A recital of the horrors experienced from violent disruptions and movements of the pack would be tiresome, for they were of almost daily recurrence. The *Terror* assumed from these enormous forces such a degree of inclination as almost threw her over on her beam ends. Finally she was squeezed and lifted entirely above the level of the sea, so that she rested on the main floe, supported only by the masses of heavy ice that had been thrown up around her by the action of the pack.

Under the influence of wind and currents the pack drifted southeastward all winter, toward Hudson Strait, the western end of which was reached in May. The winter was marked by excessively cold weather, much illness, and two deaths. The general conditions are best expressed by BACK's words, as producing 'the weariness of heart, the blank of feeling, which gets the better of the whole man. . . . No occupation, no amusement had power to please, or even to distract the thoughts.'

The release came in July. By tremendous efforts of a month's duration the surrounding ice was so cut and sawn as to secure the safety of the *Terror* when the main floe finally broke. Enormous pieces of the floe yet adhered to the bottom of the vessel, and their separation nearly capsized the ship. For four months entirely out of water, and cradled in an ice-dock, the *Terror* was so badly disabled, and now leaked so rapidly that she barely escaped foundering on her return voyage to England.

The most famous of all Arctic expeditions is undoubtedly that of Sir JOHN FRANKLIN, in the *Erebus* and *Terror*. The tragic fate that overwhelmed the entire party, the long uncertainty attending such fate, the strenuous and unavailing efforts for their relief, and the extended geographical discoveries resulting therefrom give this expedition an importance that justifies the treatment of its work and fortunes in a separate chapter.

However, years were to elapse before the discovery of the final link in the Northwest Passage by FRANKLIN was to be known by the world. Meanwhile M'CLURE was to reach in the *Investigator* M'Clure Strait, Banks Strait of PARRY, 1819, and COLLINSON to trace, in his wonderful voyage in the *Enterprise*, the winding coast-wise waters of the continent of America to the very channel in which the *Erebus* and *Terror* sank. These voyages were made, however, in attempts to relieve the missing squadron, and are therefore treated in connection with the Franklin Search by Sea (Chapter X).

ROSS, J: *Voyages for a North-West Passage* (1st, 1818; 2d, 1829–33, London 1816, 1835); PARRY: *Voyages for a North-West Passage* (1st, 1819–20; 2d, 1821–23; 3d, 1824–25, London 1821, 1824, 1826); BACK: *Voyage of the Terror* (London 1838).

CHAPTER VIII

THE NORTHWEST PASSAGE BY LAND

THE discovery of a passage across the continent of America by land, like similar ventures by sea, is inseparably associated with commercial undertakings. Land operations demanded permanent base, and this was insured by the expedition under the auspices of Prince RUPERT, which, wintering 1668–69 in Rupert River, Hudson Bay, built Fort Charles. Convinced that trade with the natives of this region would be both safe and profitable, the prince obtained from CHARLES the Second a charter, dated 2d May 1670, granting exclusive commercial privileges. This was the beginning of the Hudson Bay Company, which enjoyed these and added grants of power for two centuries. While the original charter provided primarily 'for the Discovery of a new Passage into the South Sea,' the company devoted its energies exclusively to building up a profitable trade with Indians.

Allured by Eskimo reports of rich mines of native copper, JAMES KNIGHT, governor of Nelson River, in vain urged an exploring expedition by the company until he threatened a parliamentary investigation of the charter. In 1719 they reluctantly sent out KNIGHT with two ships. The expedition never returned, and it was not until 1759 that the discovery of wreckage and ruins on Marble Island disclosed the fate of the crew, which was confirmed from Eskimo sources. Meanwhile MIDDLETON in the *California*, and MOORE in the *Furnace* discovered, in 1741, Wager Inlet, Repulse Bay, and Frozen Strait, which was followed in 1746 by a fruitless voyage inspired by the reward of 10,000 pounds offered by Parliament for the discovery of the Northwest Passage. These disastrous sea-voyages ended for a while further explorations, but in 1769, incited by repeated reports of a great river with adjacent mines of pure copper, — which seemed plausible from the pieces of metal brought in by the Indians, — the Hudson Bay Company decided to search for the source of this metallic wealth and strive for the reward for a passage.

It selected for this enterprise SAMUEL HEARNE, one of its factors at Fort Prince of Wales. His orders were to go 'far to the north of Churchill, to promote an extension of trade, as well as for the discovery of a northwest passage, copper mines, &c.' His first attempt, in November 1769, proved unsuccessful through the lukewarmness of his subordinates, the number of camp-followers, and the fact that the journey was initiated at the commencement of a subarctic winter, — when hunting on the 'barren lands' afforded precarious subsistence.

Undiscouraged by his failure, HEARNE started again 23d February 1770, from Fort Prince of Wales, with five Indians. It soon developed that it was too early for a party provided only with arms and ammunition to travel through the barren lands, where game, their only means of subsistence, was poor and scarce.

HEARNE therefore camped on Seal River, 59° 46′ N., till 27th April, when he started westward; but, obliged to travel with roving bands of Indians, — by whom he was plundered, — and having broken his astronomical instruments, he returned to Churchill 25th November. Starting twelve days later, 7th December, with MATTONABBEE, a Tinne chief, HEARNE wintered in the woods, and in the middle of April 1771, was on a small lake, 61° 30′ N., 112° W. Accumulating a stock of dried fish and meat, he turned to the north and reached Clowey Lake, 3d May. Here large numbers of Indians were assembled, and pushing on with recruits 20th June brought them to Rum Lake,— his companions venturing that far to attack the Eskimo.

Crossing the Stony Mountain they came to a (Coppermine) river, on the lower reach of which, near a cascade, they found an Eskimo encampment. Stripping to breech-clothes and moccasins, the Indians put on their war paint, and waiting till midnight fell on the sleeping Eskimo, whom, despite HEARNE's protestation, they butchered,— men, women, and children. Mutilating the bodies and destroying all property they could not carry, the Indians accompanied HEARNE to the mouth of the Coppermine, which was reached 17th July 1771. The water was fresh, but seals, ice, and other signs convinced him that from the islet-strewn delta he overlooked the northern ocean.

HEARNE's claim to have reached the polar sea was very seriously questioned, for he made no astronomical observation, but located it by dead reckoning. FRANKLIN's journey of 1821 confirmed HEARNE's claims, through his accurate description of the cascade, where unburied skulls yet marked the massacre. HEARNE very greatly over-estimated distances, and placed the mouth of the Coppermine 4° of latitude too far north and 5° of longitude to the west, the true position being about 67° 48′ N., 115° 47′ W.

A spirit of adventure and exploration instigated the next journey across the barren lands, through warring tribes and down torrential rivers to the Arctic coast of northwest America.

One of the Northwest Fur Company, stationed at Fort Chipewyan, and impressed by Indian accounts, believed that the northern sea could be reached by a large river issuing from Great Slave Lake. This man was ALEXANDER MACKENZIE, — afterward Sir ALEXANDER, — made famous by this and a later journey across British America, — the first transcontinental passage made north of Mexico.

With a number of canoes MACKENZIE left Fort Chipewyan, 3d June 1789, and reached Slave Lake via Peace River, — a well known route. Delayed by ice and making trading arrangements with the Indians, it was 25th June when, with a new guide, he started in search of the outlet of the (Mackenzie) river, which was found on the 29th. Undeterred by Indian tales of rapids, monsters, and Eskimo, he boldly intrusted himself to this unknown stream, obtaining guides by persuasion, pay, and force as occasion required. Adversities and dangers were encountered, but it is of interest to note that the hardship of which MACKENZIE complains most in his journal was attacks of clouds of mosquitoes. On 12th July 1798, he arrived at Whale Island, in the river delta, 69° 14′ N. by observation, 135° W. by dead reckoning. Sea-fowl, whales, and a dangerous swell indicated that he was on the shores of the polar sea, which was not salt, for in that almost tideless region, the great volume of the Mackenzie keeps the water always fresh many miles seaward.

The successor of MACKENZIE was the best-known man connected with the Northwest Passage, whether by land or by sea, — Sir JOHN FRANKLIN. A midshipman in the leading ship in the battle of Copenhagen, distinguished for intrepidity at Trafalgar, one of the few English officers who gained laurels at New Orleans, wrecked on an Australian coral reef, and serving successfully seven years as Governor of Tasmania, yet his fame and fate were linked rather with Arctic exploration than with naval service or civil duties. Associated as commander of the *Trent* with BUCHAN in 1818 (Chapter XII), he sought Arctic service at the first opportunity, which came the next year. The Admiralty Board, in renewing Arctic exploration in 1819, decided to make attempts by land and by sea. It sent PARRY by sea, while it fell to FRANKLIN's lot to outline the north coast of America, where HEARNE and MACKENZIE had located two isolated points, the mouths of the Coppermine and the Mackenzie.

FRANKLIN's work was to be done by sledge and canoe, through the Hudson Bay territory, with Dr JOHN RICHARDSON, midshipmen ROBERT HOOD and GEORGE BACK, and a seaman, JOHN HEPBURN. A dangerous voyage in one of the company's ships brought them to York Factory 30th August 1819. An ordinary man would have tarried there for the winter, but FRANKLIN with his usual energy pushed on and completed by 22d October an autumnal journey of 700 miles, involving marches, portages, and rapids, that brought them to Cumberland House.

FRANKLIN's ambitious spirit was not satisfied with this measure of success, but on 18th January 1820, in the middle of an Arctic winter of intense cold and prolonged darkness, with BACK and HEPBURN he started northward for Fort Chipewyan, Lake Athabasca. The two months' journey of 800 miles involved terrible hardships. The supplies were hauled on dog-sledges, while the men travelled over the snow-covered land on snow-shoes, which, as they were *mangeurs de lard*, or novices, caused almost intolerable pain in their swollen feet as they strove to keep pace with the dogs. The temperature fell as low as 70° below the freezing-point, and occasional blizzards barely failed of destroying the whole party.

FRANKLIN, unable to obtain sufficient supplies at Chipewyan, pushed on to Fort Providence, Great Slave Lake, where on leaving he had only ten days' provisions for his party. The entire force now consisted of 26 men, three Indian women and children, with whom FRANKLIN hoped to reach that autumn the lower Coppermine, his proposed winter-quarters. The journey lay up the Yellow Knife. The stock of food soon failed, fish and game were scarce, portages frequent, long, and difficult, signs of coming winter appeared, and much against his will FRANKLIN was constrained, after a journey of 553 miles, to build on Winter Lake, Fort Enterprise, 64° 28′ N., 113° W. This display of energy by FRANKLIN was counteracted largely by the neglect and misconduct of the officers of the Hudson Bay and Northwest companies, who refused to forward the promised supplies, and were believed to have bred among the Indians distrust of FRANKLIN's intentions.

BACK volunteered to return to Providence and Chipewyan for supplies absolutely necessary for the success of the expedition, — ammunition in particular, which was almost exhausted. Meanwhile about 180 deer were killed and over 1,200 fish caught, but the Indians crowded to the fort in such numbers as threatened to exhaust the food during the winter, when game was absent, so FRANKLIN was obliged to dismiss some of them. The winter proved to be intensely cold, and FRANKLIN says, 'The trees froze to their very centres and became as hard as stones, and more difficult. Some of the axes were broken daily.' More than once the temperature of the bedroom before fire was made was as low as 70° below freezing.

BACK returned with ammunition and other supplies the 17th of March, having travelled over 1,100 miles on snow-shoes, with no covering at night beyond a blanket and deerskin, sometimes from two to three days without food, and in temperatures often 70° and once 90° below freezing. The conditions of this mid-winter journey may be surmised by BELANGER's appearance on arrival at Enterprise: 'His locks were matted with snow, and he was encrusted with ice from head to foot, so that we scarcely recognized him as he burst in on us.'

It was 1st July 1821 before FRANKLIN was able to accumulate men and supplies on the banks of the Coppermine, but on the 18th, after numberless hazards, the ocean was reached at the mouth of the Coppermine, — 67° 48′ N., 107° W., — 350 miles from Fort Enterprise. FRANKLIN says: 'The position differs widely from that assigned by Mr HEARNE; but the accuracy of his description, conjoined with Indian information, assures us that we were at the very part he visited.' Provisions for 15 days only remained, and the immediate country was absolutely barren; nevertheless FRANKLIN decided to explore the sea-coast to the eastward. Bands of Eskimo met near the sea proved friendly, and FRANKLIN counted on aid from other bands that frequent the coast to the eastward.

His *voyageurs*, at home in rapids or in forest, where they faced unhesitatingly the dangers of the march or chase, now viewed with terror a boat voyage through an icy sea. The length of the journey, the roughness of the waves, the uncertainty of provisions, the absence of fuel created such apprehensions that they sought their discharge, and were with difficulty shamed into proceeding.

A new country lay before FRANKLIN, who, starting on 21st July, by indefatigable exertions succeeded in tracing the south shore of an extensive sound (Coronation Gulf), the entire coast of an inlet (Bathurst), and finally attained Point Turnagain, near a (Dease) strait, 68° 18′ N., 109° 25′ W., on 16th August 1821.

The journey through an ice-encumbered sea, along a coast mostly rocky and high, had by this time so injured his two canoes that they were daily in danger of falling to pieces. But three days' pemmican remained; his men were most apprehensive of their safety; the Eskimo assistance expected failed, although frequent traces were found along the whole coast; indeed progress to this extent had only been possible by the occasional deer, bear, and other game that was secured in intervals of travel. Most reluctantly FRANKLIN turned back on 22d August, the same day that PARRY (Chapter VII) sailed out of Repulse Bay, 539 miles distant.

The freezing weather had already set in, with snow, which, with the insufficiency of food and the scarcity of game, made it impossible to return over his outward route. FRANKLIN consequently determined to enter Hood River, one of his outward discoveries, and from the head of canoe navigation cross overland to the Coppermine and Fort Enterprise. The hope of successful navigation in Hood River was destroyed by reaching impassable rapids. The river fell 250 feet in a mile, interrupted by two falls 60 and 100 feet high respectively, in its passage through a long canyon with vertical walls of 200 feet. They were now 150 miles distant from Point Lake, for which they started on foot, the *voyageurs* carrying two small canoes, made out of the remains of their boats. Starting on 31st August they ate that day their last food, except a little portable soup and arrow-root. On 4th September they were delayed three days by an extremely violent storm, which kept them in camp without food or fuel, in a temperature far below freezing. On the 7th FRANKLIN fainted at starting, and the last of their portable soup and arrow-root, eaten after a three days' fast, was cooked from a fire made from one of their canoes, which was broken,— possibly by an improvident *voyageur* to avoid carrying it. The snow a foot deep, the country marshy, and the new ice breaking and plunging them into icy water, made their journey one of indescribable difficulty and discomfort.

The main party travelled in single file, the officers

breaking the way through the deep snow, while the hunters, keeping well on the flanks, scoured the desolate country for needed game. Little was found, and for days at a time they lived on *tripe de roche* (lichens), deer-bones left by wolves, etc., — occasionally having part of a partridge or a few berries. One day word came to FRANKLIN that the remaining canoe was so broken as to be useless and had been abandoned. With anguish he realized that it was practically their death-warrant, and besought the two *voyageurs* to return and bring it up as it was. The ignorant obstinacy of the men caused them to refuse, and as no officer could carry the canoe they went on without. One hunter straying perished; otherwise the party reached the Coppermine, — which all knew to be a rapid, unfordable stream.

They were now within 40 miles of Fort Enterprise, which could easily have been reached by all in the eight days taken to cross the Coppermine. FRANKLIN'S advice was ignored and the obstinate *voyageurs* fruitlessly wasted their time in seeking fords, searching for timber, and in the construction of willow rafts. These last nearly caused the death of RICHARDSON, who attempted unavailingly, under heroic and desperate conditions, to swim the river to get over a line by which the raft could cross. Eventually they came to FRANKLIN'S idea, a willow-framed boat covered with the canvas bedding; which proved practicable, — the party crossing on 4th October.

BACK and two men were sent ahead to Fort Enterprise, to secure aid for the party, for it was evident to FRANKLIN that some were near the end of their strength.

Tripe de roche (lichens) and scraps of roasted leather now formed their only diet, which disagreeing with two men they fell down with exhaustion on the second march, 6th October, in the midst of a strong gale and drifting snow that drained their remaining vitality. Despite the generous and heroic efforts of RICHARDSON and HOOD they perished where they fell. These two officers, with HEPBURN, who volunteered, established a relief camp at the first place where enough *tripe de roche* and wood for ten days could be found. FRANKLIN reluctantly consenting to this arrangement, pushed on with the eight remaining men. Four of them, however, eventually exhausted, returned to join RICHARDSON, — MICHEL, a hunter, being of the number. Two of the *voyageurs* had been detected in stealing food since leaving Hood River, and the hunters had systematically appropriated to their own use small game shot away from the main party. This system of misappropriation, fatal to any party if not sternly dealt with, proved destructive to this expedition. MICHEL, actuated solely by the desire to save his own life at the expense of all others, killed three *voyageurs* and Lieutenant HOOD. RICHARDSON, as a matter of preservation of HEPBURN and himself, shot MICHEL.

In the meantime FRANKLIN with four men reached Fort Enterprise and found it entirely abandoned. A note from BACK two days old, told him he was pushing on for assistance to Fort Providence, about two weeks further. Joined later by RICHARDSON and HEPBURN, FRANKLIN'S small party subsisted on the bones and skins of deer, remaining from the previous year, *tripe de roche*, moss, etc. Two men eventually died, and the whole party would have perished had it not been for the successful efforts of BACK, who, falling in with Indians a few days' march from Fort Enterprise, obtained such assistance from their chief, AKIATCHO, as assured safety for the remainder of the party. BACK in his trip lost one man from cold and starvation.

On FRANKLIN'S return to England, the British government, having determined to again send PARRY to effect the

Northwest Passage by sea, approved of a plan for an expedition overland to the mouth of the Mackenzie, and thence by sea to the northwest extremity of America in order to survey the unknown coast. The plan was FRANKLIN'S, and to him was intrusted its execution. He left Great Britain in 1825, at a time when his wife was upon her death-bed, she, however, insisting that her husband should proceed upon this journey despite her condition. Dr RICHARDSON and Lieutenants BACK and HEPBURN accompanied him. Their winter quarters were fixed on Great Bear Lake, not only on account of its proximity to the mouth of the Mackenzie, but from its abundant store of fish. Here, in latitude 65° 12′ N., longitude 123° 13′ W., Fort Franklin was erected.

The plan contemplated RICHARDSON, with a boat party, examining the coast between the Mackenzie and the Coppermine rivers, while FRANKLIN was to proceed westward from the mouth of the Mackenzie, skirt the North American shore to Icy Cape, — the farthest of COOK in 1777, — and reach, if possible, Kotzebue Inlet, where BEECHEY, in the *Blossom*, was to meet him.

Before establishing his post on Bear Lake, FRANKLIN made a visit to the mouth of the Mackenzie from Fort Norman (64° 40′ N., 124° 53′ W.), situated at the confluence of that stream with the River of the Mountains. Leaving Fort Norman on 8th August 1825, he reached the mouth of the Mackenzie without serious difficulty on the 18th, encamping in 69° 29′ N., 165° 41′ W. The returning party reached Fort Franklin on Great Bear Lake 5th September, which had been built for them.

The winter of 1825–26 was spent in hunting, fishing, and in making a series of meteorological, magnetic, and other physical observations. Game was fairly abundant, despite which the Indians in mid-winter frequented the fort in a state of semi-starvation due to their indolence. During April RICHARDSON completed the survey of Great Bear Lake; preparations for the main work of exploration began in June.

FRANKLIN left his winter quarters 22d June 1826, and reached the delta of the Mackenzie 3d July, where he and RICHARDSON were to part, each with two boats and provisions for eighty days. FRANKLIN'S party consisted of 16 men all told; BACK commanded the second boat, with an equal number of men.

At the mouth of the Mackenzie FRANKLIN approached an Eskimo camp to open friendly relations with it, but the water was so shallow that the boats grounded far from shore. The natives came out in kayaks, peaceful advances were made through the interpreter, and trade commenced. With the ebbing tide the natives surrounded the stranded boats and despite the remonstrances of their own chiefs endeavored to plunder them. Emboldened by FRANKLIN'S mildness they even attempted to disarm him and his men, and nearly overpowered his crew. Fortunately BACK'S boat got afloat and that officer judiciously ordered his men to level their muskets, which fortunately scared off the Eskimo and ended the affair.

Herschel Island, 69° 34′ N., 140° 51′ W., reached the 17th July, proved to be inhabited by Eskimo. Ice and fog made farther progress slow, and 5th August found them at Flaxman Island, 70° 11′ N., 145° 50′ W. Detained nearly a week by fog at a small island, FRANKLIN succeeded in reaching, 16th August 1826, a small (Return) reef, which was in 70° 26′ N., 148° 52′ W. Here a gale sprang up, and the prospect of reaching Point Barrow, before discouraging, was now hopeless.

The situation was complicated by the fact that Captain F. W. BEECHEY, in the *Blossom*, was ordered to meet

FRANKLIN in Bering Strait, and bring his party back to England. As Icy Cape had never been doubled, FRANKLIN was justified in his opinion that BEECHEY would not pass Kotzebue Sound,—which was true as far as anchorage of the *Blossom* was concerned. BEECHEY, however, was impressed with the importance of pushing along the coast as far as possible to the eastward, for he well knew FRANKLIN'S determined spirit. With this view BEECHEY, unable to reach land north of Cape Lisburne, started northward Mr ELSON, with a barge, 17th August,—while FRANKLIN was encamped at his farthest, Return Reef. This energetic officer succeeded, despite unfavorable ice-conditions, in rounding Icy Cape, and traced a new and unknown coast to the most northerly mainland in the Bering Strait region,—Point Barrow, 71° 24′ N. ELSON reached this point, which is within 160 miles of Return Reef, 22d August, and, ascertaining from the natives that no white men had visited them, started back to the *Blossom*, in Kotzebue Sound, three days later.

Meanwhile FRANKLIN very wisely decided to turn back to Mackenzie River, 374 miles distant, which was reached 29th August 1826. If FRANKLIN had pressed on to the west it is evident that he could not have caught ELSON, and the future of his party would have been most problematical, for the following year, 1827, the *Blossom* was unable to get within 100 miles of Icy Cape.

FRANKLIN found Dr RICHARDSON safely returned to Fort Franklin from a most successful voyage. RICHARDSON left the Mackenzie by the most easterly of its navigable mouths, and there fell in with a band of Eskimo that were friendly until his boats grounded, when they attempted to capture them, and desisted only on arms being levelled. Proceeding northeast along the coast, RICHARDSON landed, 13th July, at Atkinson Island, where he found 17 winter-houses and a large public building, the latter 27 feet square with log roof.

Ice, fog, and storms delayed them; on such occasions hunting, fishing, and collecting plants, etc., utilized their time. It was not until 18th July that they rounded a cape (Bathurst), 70° 36′ N., 127° 35′ W., the most northerly point of their voyage, and, with the exception of Barrow and Boothia, of the American continent. Their course now lay to the southeast across Franklin Bay and round Cape Parry, a rocky promontory 700 feet high, from which, 23d July, they saw land to the south-southeast, 40 miles distant. Clerk Island was passed 1st August and they were now entering a (Dolphin and Union) strait, — named for their boats. The prevalent fog prevented their ascertaining this fact until they reached Cape Hope, 68° 58′ N., whence they saw land to the north, distant some twelve miles, with an intervening strait densely packed with ice. The land, named Wollaston, was moderately high, separated from the American continent by a (Dolphin) strait from twelve to 15 miles wide. As they rounded Cape Krusenstern and entered Coronation Gulf, Wollaston Land extended continuously to the east.

RICHARDSON was safe at the mouth of the Coppermine 8th August, having traced the continental coast-line through 20 degrees of longitude and two of latitude, discovered a new land and strait, determined the tidal conditions of the polar sea, and thoroughly examined the geological formations of the Arctic coast and made collections of its flora, — of which he obtained 170 flowering species. Fort Franklin was reached 1st September 1826, after a journey, by land and sea, of 1,709 geographical miles.

In 1832 the protracted absence of Captain JOHN ROSS, since 1829, properly excited grave apprehensions for the safety of his party. The government contributed $10,000 for an expedition under BACK, via Great Fish or Back River, while friends of ROSS contributed $15,000; the Hudson Bay Company provided canoes and supplies.

Accompanied by Dr RICHARD KING and three men, BACK reached Fort Resolution, 8th August 1833, — a post of the Hudson Bay Company that was to be his main base of operations. Appreciating the importance of a preliminary examination of his route for the coming spring, — especially as it had never been travelled over, — BACK commenced operations at once. With very great difficulty the journey to the Great Fish (Back) River was made over a series of small lakes and rivers, — interrupted by many portages — from the east end of Great Slave Lake. The illness of his interpreter, the desertion of two Indians and the attempt of another to abandon him, did not deter BACK from continuing his advance. The search over a rugged, unknown country for a river indefinitely located by Indian reports, presented peculiar difficulties, but BACK, on 31st August 1833, had the satisfaction of reaching the Great Fish River, in 64° 41′ N., 108° W. The lateness of the season forbade extensive exploration of the new-found river, — the Great Fish, or better, Back River, — and they returned to Great Slave Lake, after proceeding as far as the first rapids.

During his absence progress was made in the construction of winter-quarters, which were built at the eastern extremity of Great Slave Lake, and named Fort Reliance, about 63° N., 109° W. The winter was unusually cold, following a warm autumn, and deer were very scarce. In consequence BACK had to see the Indians who congregated around Fort Reliance suffer indescribably from hunger and exposure, that killed many.

The spring of 1834 brought news of the safe return of ROSS and his party. The particulars confirmed, as BACK says, 'the wisdom as well as the humanity of the course assumed by the promoters of our expedition.'

While the principal object of his journey had been unexpectedly attained, yet BACK felt that he must carry out the geographical and scientific work in view. The spring explorations were attempted with ten men, and their single boat was launched in Back River, 27th June 1834. Dangerous, rocky rapids, ice-covered lakes, — enlargements of the main stream, — and impassable cascades and falls, entailing laborious portages, were the various phases of this dangerous journey.

One experience in an exceedingly dangerous piece of rocky rapids is the origin of a well-known but unplaced story. When a snapping oar left it for a moment uncertain whether their boat would not be dashed stern foremost on the sunken rocks, in the instant of suspense 'one of the crew, with less nerve than his companions, began to cry aloud for aid. M'KAY, in a still louder voice, exclaimed, "Is this a time for praying? Pull your starboard oar."'

They fell in with Eskimo 28th July 1834, and two days later were at the mouth of the Great Fish River (Back), 67° 11′ N., 94° 30′ W. They had run 530 geographical miles down a rock-bound river, experiencing the dangers of no less than 83 cascades, falls, and rapids.

The new discoveries of BACK were destined to be confined to this dangerous river and the barren lands adjacent to its mouth. The sea-ice was packed heavily in the mouth of Back River, and with great difficulty BACK passed an island (Montreal) and reached a point (Ogle), where he made his farthest camp 15th August, in 68° 14′ N., 94° 58′ W. This was at the eastern entrance of a strait (Simpson), — afterward passed by DEASE and SIMPSON, 1839, — the greater part of which was seen by a party of BACK's that walked some 15 miles to the west of Point

Ogle. This western passage and the finding of a large log caused BACK to believe, rightly, that there was free water communication to the westward rivers. To the northward was a new land, and another point had been established on the unknown coast of the continent of America.

A project of sending a party to the west toward Point Turnagain of FRANKLIN, proved impracticable, for the country was so boggy that a loaded man sank to his knees at every step. Naming after King WILLIAM the new land — destined to be forever associated with the fate of FRANKLIN — BACK turned southward, repassed the rapids, wintered at Reliance, and returned to England.

In 1836 the Hudson Bay Company decided to send out an expedition 'to endeavor to complete the discovery and survey of the northern shores of the American continent.' At this time to the westward, there was an unknown shore of 150 miles or more between Return Reef of FRANKLIN and Point Barrow of ELSON; to the eastward between Point Turnagain of FRANKLIN and the Melville Peninsula of PARRY, the only known portions were the discoveries of Sir JOHN ROSS in Boothia Land and those of Sir GEORGE BACK at the mouth of Back, or Great Fish River.

The undertaking was viewed by many as impracticable; but among the many energetic and capable employees of the Hudson Bay Company were found two men, whose capacities and judgment were deemed equal to the undertaking. These men were P. W. DEASE and THOMAS SIMPSON, — DEASE the older, the more experienced, but the latter a man of great ambition and singular resolution, to whose personal exertions may be attributed the wonderful results that flowed from this expedition. They may be well called wonderful, for SIMPSON succeeded in reaching Point Barrow to the west, and the westerly shore of King William Land to the east. Thus by overlapping the discoveries of BEECHEY in one direction and the later route of Sir JOHN FRANKLIN in the other, SIMPSON directly established in conjunction with these two the existence of a northwest passage by water. SIMPSON, having visited Fort Garry to refresh his astronomical studies, was obliged to make a midwinter journey to rejoin DEASE, who wintered at Fort Chipewyan. Starting 1st December, SIMPSON fixed on 1st February as the date for reaching Chipewyan, 1,277 miles distant, and arrived on the very day. He was accompanied from post to post of the Hudson Bay Company by fresh men and dogs, but he says 'I had myself raised — (that is, broke the path through the snow) — the road through the whole journey, my companions being sufficiently occupied, each with the care of his sledge.' This journey of SIMPSON, and those of FRANKLIN and BACK in the same inhospitable region, are remarkable exploits, on account of their accompaniments of blizzards, cold, etc.

DEASE and SIMPSON left Chipewyan 1st June 1837, and delayed by ice ten days at Fort Resolution reached the mouth of the Mackenzie and entered the polar sea 9th July. Every day or two they met parties of Eskimo, who proved always friendly. Fog, storm, and ice were experienced; nevertheless they reached Return Reef, of FRANKLIN, 23d July. SIMPSON says, 'Our early arrival at the point where our discoveries were to commence is, under Providence, mainly attributable to our inflexible perseverance in *doubling* these great ice-packs, any one of which might have confined us a fortnight to the beach.'

The same line of action was pursued in the westward voyage, during which the party once ventured some 17 miles from land and were nearly caught in the great pack. At Cape Simpson the ice was so thick that four days were spent in making as many miles. Finding it impracticable to reach Point Barrow by boat, SIMPSON determined to proceed there overland from this point, 71° 03′ N., 154° 26′ W., it being only two degrees to the westward. DEASE consenting to care for the boats, SIMPSON and five men started 1st August on foot, carrying on their backs a canvas canoe, arms, food, etc., — about 45 pounds to a man. Two marches brought them to an Eskimo camp, where they most fortunately obtained three large skin-boats and rowers for the remainder of the journey to Point Barrow, which was reached 4th August. SIMPSON could well say, 'Landing I saw with *indescribable emotions* Point Barrow, stretching out to the northward,' for he had completed the northwestern coast-lines of the continent of America. Determining the position of the point, 71° 24′ N., 156° 20′ W., SIMPSON left the next day, and made a rapid and successful journey homeward.

Wintering at Fort Confidence, 66° 54′ N., 118° 49′ W., which was built for their winter quarters near the mouth of Dease River, the party subsisted almost entirely by hunting and fishing. The Dease River region was explored so that suitable portages for crossing to the Coppermine could be laid out. On 6th June 1839, the start was made, and partly by water and partly on sledges the boats were transported via Kendall River to the Coppermine.

On their arrival, 25th June, the Coppermine had not yet broken up, and with delays from ice and high water they did not reach the sea until 1st July. The time however, was well-occupied in hunting, which materially increased their stock of provisions. Sixteen days elapsed before the sea ice permitted their departure, and then the ice conditions in the bay — into which the Coppermine empties — were so bad that they did not reach Cape Flinders, 68° 16′ N., 109° 21′ W., until 9th August. Here several gales detained them ten days, with no water in sight. In this contingency SIMPSON resorted to his old tactics, and 20th August started eastward on foot, with seven men, carrying tent, canvas canoe, arms, and provisions for ten days. DEASE was to follow with boat when practicable. Point Turnagain, of FRANKLIN, 1821, was passed the first day and SIMPSON's discoveries began. He covered 100 miles of new coast, reaching, 24th August, 68° 44′ N., 106° 03′ W. From a bold headland, Cape Alexander, 68° 52′ N., — his most northerly point, — SIMPSON saw that he was at the eastern entrance of an ice-obstructed (Dease) strait. To the north rose an extensive unknown coast, the easterly extension of Wollaston Land, of RICHARDSON, 1826, to which SIMPSON gave the name of Victoria. Exhausted but exultant, the party rejoined DEASE 29th August 1838, and returning again wintered at Fort Confidence.

The 22d of June 1839 found SIMPSON and DEASE at the mouth of the Coppermine, where the sea was yet solid. Exploring Richardson River during this delay, they put to sea at the earliest opening, and favored by ice conditions reached by boat Cape Alexander, 26th July, when the sea to the east was found closed. Working east as ice openings permitted, they skirted the southern shores of Victoria Strait, and discovering 11th August a new (Simpson) strait reached its eastern entrance, Point Seaforth, 68° 32′ N., 97° 35′ W. Point Ogle, of Sir GEORGE BACK, 1834, was attained 13th August, and two days later SIMPSON visited Montreal Island, — where he found in bad order the food cached by BACK five years before.

This accomplished their instructions, but SIMPSON pushed on to the east and reached, 19th August 1839, their farthest, Castor and Pollux Bay, 68° 28′ N., 94° 14′ W. Turning homeward the next day, they kept to the north shore of Simpson Strait, erroneously thinking this (King William) land to be Boothia Felix of JOHN ROSS, 1829. Tracing its coast some 60 miles they reached a cape (Herschel) 68° 41′ N., 98° 22′ W., 57 miles from JAMES C. ROSS's Pillar, and within 90 miles of the north magnetic pole.

As only nine years later the Franklin expedition perished from starvation on this very land, it is interesting to note that SIMPSON reports it in 1839 as 'a country, abounding in reindeer, musk-cattle, and old native encampments.'

Crossing to the west, SIMPSON later coasted along the south shore of Victoria Land from east of Cape Colborne to Cape Peel, which is west of an inlet (Cambridge Bay) where COLLINSON wintered in the *Enterprise*, 1852–53. Altogether they traced 156 geographic miles of this land (Victoria), which SIMPSON thought to be separated from Wollaston Land to the west, — an erroneous surmise, as the explorations of COLLINSON showed in after years.

The party reached the mouth of the Coppermine, 16th September 1839, after an Arctic boat journey of 1,408 geographic miles, the longest on record.

All the lands visited by SIMPSON were dotted here and there by Eskimo encampments, — some occupied, some temporarily vacant, and others long since deserted. Game was abundant, both on land and at sea. However, theory teaches, and experience confirms theory, that game in the Arctic regions is especially migratory, from a variety of causes, so that the expectation of subsisting an expedition on the game of any section rests on a dangerous fallacy.

The successful return of this expedition closed the remarkably successful career as an explorer of THOMAS SIMPSON, who died by violence the following spring while awaiting orders to assume command of another expedition.

While SIMPSON's original discoveries did not equal in extent those of FRANKLIN, yet profiting by the experiences and charts of his great predecessor, SIMPSON's journeys far exceeded FRANKLIN's in length and duration. SIMPSON and

RAE, it may be here said, worked in a country and under conditions with which they had for years been familiar, while FRANKLIN, BACK, and RICHARDSON were obliged to supply their practical ignorance of conditions and means by fertility of resource and ability of application.

The great success of DEASE and SIMPSON, 1838-39, caused the Hudson Bay Company to arrange for another expedition in 1840, which was to have been commanded by SIMPSON, but his untimely death led to the abandonment of the voyage. A few years later Sir GEORGE SIMPSON, then governor of Hudson Bay, planned an expedition which was to have Repulse Bay as its base, from which point it was believed that Boothia Felix, and the rest of the unknown coast of the continent could be surveyed. Fortunately for its success, the expedition was put under the command of Dr JOHN RAE, whose natural abilities, fine physique, conjoined to an experience of ten years in the Hudson Bay territory, admirably fitted him for the task.

Wintering at York Factory, 1845-46, RAE, with two boats and ten men, left that post 11th June 1846, and reached Churchill Station 16 days later. His letter of instructions, received a week later, laid on him an amount of scientific duties that would require a corps of assistants at the present day, and, besides assigning him to command, enjoined him to personally attend to subordinate duties. He was to determine astronomically all remarkable points, make bearings of all intermediate portions of the coast, chart these daily, 'attend to botany and geology; to zoölogy in all its departments; to the temperature of air and water,' to the atmosphere, ice, winds, currents, soundings, magnetic dip and inclinations, aurora borealis, refraction of light, ethnographic peculiarities of the Eskimo, and such 'other (observations) as may suggest themselves to you,' etc. It is evident that the Hudson Bay Company wished full returns for its money.

It should moreover be borne in mind that Rae was sent into this unknown country with only four months' full rations, although the plan contemplated his absence for either 15 or 27 months. In case the game resources of the country would not support all his men he was to send back some; and the letter coolly recited that he could not fail to find subsistence for the rest of his men, animated as he was to fulfil his 'mission at the cost of danger, fatigue, and privation.'

Taking two Eskimo as interpreters, Rae left Churchill 5th July 1846, the day after receiving his orders, and used such despatch that he reached the head of Repulse Bay 25th July. Meeting Eskimo there he at once had them draw a chart and give him information regarding the country to the north. They told him that the bay to the west of the Melville peninsula was distant overland but 40 miles, and that the route lay across a low country, where a chain of lakes left only twelve miles of portages.

Rae commenced operations that very day, and found the statements of the natives correct. The tide-water of Committee Bay was reached 1st August, in 67° 13′ N. Here other natives were fallen in with, and an Eskimo woman drew a chart from which Rae was led to believe that the only outlets to the bay were Regent Inlet to the north and Fury Strait to the east. It seems strange that John Ross, 1829–33, should have passed Bellot Strait, the western exit, and that he in common with Rae should have met Eskimo who could give no information concerning this final link in the Northwest Passage.

Following the west side of Committee Bay, Rae was ice-bound in 67° 30′, and returning south was no more fortunate on the east side. Turning homeward he crossed the (Rae) isthmus with three men on foot, making the 50 miles in two days over ground so rough that they reached Repulse Bay 'rather foot-sore, our shoes and socks having been entirely worn through long before our arrival.'

Preparations for winter quarters were immediately made. Part of the men collected stones for a house, while others set nets, gathered fuel, moss, and heather, or hunted. The stone walls of their house were roofed with boat-sails, and the door was made of parchment deer-skins. The house, called Fort Hope, was in 66° 32′ N., 86° 56′ W. When game failed in November, they had already killed 144 deer, 14 hares, 180 ptarmigan, and caught 200 fish. Observatories, store houses, etc., were built out of snow, and united by snow galleries. The natives who wintered and hunted near Fort Hope occasionally sold them oil for fuel, but this source failing late in the winter, they were so reduced in this respect that they took for many weeks only one meal a day. During a severe gale the temperature in their house fell to zero or below; and one man had his knee frozen in bed, and 'letting it be known, got heartily laughed at for his effeminacy.'

The end of February 1847 brought back deer, but none were shot, owing to their shyness, till the middle of March. Sledges were made, dogs bought, and other preparations made for the spring sledging.

On 5th April 1847, RAE started on his sledge journey, with five men, two sledges, and eight dogs. The Eskimo neighbors helped him over Rae Isthmus, and then his hard travel began along the west shore of Committee Bay. Two dogs died, and the work told severely on all. On 16th April, leaving half his party in 68° 54′ N. to hunt, and all his dogs to recruit, he proceeded with three men on foot. Three days later RAE reached a high divide, 69° 31′ N., 91° 30′ W., overlooking Lord Mayor Bay of JOHN ROSS, 1829–33, thus connecting Hudson Bay with the northern discoveries of PARRY and ROSS.

On reaching Fort Hope after their journey of nearly

600 miles, RAE says, 'We were all well, but so black and scarred on the face from the effects of oil, smoke, and frost-bites, that our friends believed we had met with some serious accident from explosion of gunpowder.'

On 12th May, RAE started again in order to follow the east shore of Committee Bay to Fury and Hecla Strait. A supporting party with a dog team accompanied him three days, after which he proceeded on foot with four men. Progress was very slow and provisions were scanty, but fortunately RAE shot a deer and they reached, 27th May 1847, 69° 20′ N., 85° W. RAE walked five miles farther north, whence, from a headland, he was able to see twelve miles to Cape Ellice, ten miles of the south of Fury Strait (PARRY, 12th September 1822).

Fort Hope was reached 9th June 1847, and the return boat journey was made to York Factory between 12th August and 6th September. The number of scientific observations made and recorded by RAE in this journey must be considered as remarkable, and these with his collections indicate how much can be done by a man of RAE's untiring energy and uncompromising fidelity.

This expedition practically completed the exploration of the coast of North America. There remained only the northwest shores of Boothia Felix, and the southeastern part of King William Land to make the survey complete. These were destined to be filled in during the Franklin Search, and in connection therewith they are treated.

During his service, 1863–1878, in Arctic America, a French priest, M. ÉMILE PETITOT, made many journeys in the Mackenzie basin and around the Great Bear and Great Slave lakes. His maps and publications are important geographic and ethnographic contributions on these comparatively unknown regions.

Other earnest priests and missionaries have devoted themselves to the Christianizing of Arctic America from

Alaska eastward to the Mackenzie basin. The immigrant and miner are gradually occupying these remote regions, and even the pulse and throb of steam are not wanting on the lower Mackenzie, and along the coast to the northwest.

The waning whaling industry, which from 1874 to 1890 in the waters adjacent to Bering Strait, made a catch aggregating over eleven million dollars in value, is steadily exploiting its last fishing grounds, between Point Barrow and Banks Land. With a permanent relief station at Point Barrow, the ships of the Pacific Whaling Company find it safe and profitable to winter at Herschel Island, near the mouth of the Mackenzie, and in 1889 Commander STOCKTON carried the U. S. S. *Thetis* to the Mackenzie, and thence by a most successful voyage passed west to Wrangel Island.

Thus the coasts and regions made known through the daring explorations of MACKENZIE, FRANKLIN and SIMPSON, have by broadening commercial and industrial industries arisen from their position of unknown, worthless wastes, to the dignity of regions contributory to the needs and desires of man.

HEARNE: *Journey to Northern Ocean* (London 1795); MACKENZIE: *Voyages to Frozen and Pacific Oceans* (London 1801); FRANKLIN: (First & Second) *Journey to Shores of Polar Ocean* (London 1824, 1828); BACK: *Arctic Land Expedition* (London 1836); KING: *Journey to Arctic Ocean* (London 1847); SIMPSON: *Discoveries on the North Coast of America* (London 1843); RAE: *Expedition to the Arctic Sea* (London 1850); PETITOT: *Les Grandes Esquimaux; Grand Lac des Esclaves; Grand Lac des Ours* (Paris 1887, 1891, 1893).

CHAPTER IX

FRANKLIN'S LAST VOYAGE

THE agitation of the Royal Geographical Society for farther exploration of the northern coast of America and search for the Northwest Passage, was successfully renewed when the return of JAMES C. ROSS from the Antarctic seas left two well-found ships, the *Erebus* and the *Terror*, available. Captain JOHN FRANKLIN, just returned from seven years of service as governor of Tasmania, was 59 years old, but as full of Arctic enthusiasm as ever. He would not apply for the duty, but when all turned to him as fitted by experience and capabilities for this dangerous service, he accepted unhesitatingly. He said: 'No service is dearer to my heart than the completion of the survey of the northern coast of America and the accomplishment of the Northwest Passage.'

With FRANKLIN sailed CROZIER, who had served with PARRY in 1821, 1824 and 1827, and with Sir JAMES C. ROSS shared the honor of having approached the nearest to both the geographic poles. GORE had served with ROSS in the *Antarctic*, and with BACK in the *Terror*, while other officers had equally distinguished themselves on duties demanding courage, ability, and knowledge. The ships were put in the best possible order, and everything was done that promised to either insure the safety and success of the expedition or promote the health and comfort of the men. The plan followed was that of Sir JOHN BARROW, and to make clear the situation, both as regards FRANKLIN'S last voyage and also as to the methods followed for his relief (Chapters X and XI), part of the official instructions, dated 5th May 1845, are reproduced:

'SECTION 5. Lancaster Sound and its continuation, . . . having been four times navigated without any impediment by Sir EDWARD PARRY, . . . will probably be found without any obstacles. In proceeding to the westward, you will not stop to examine any openings either to the *northward* or southward in that strait, but continue to push to the westward without loss of time, in the latitude of about 74¼°, till you have reached the longitude of that portion of land on which Cape Walker is situated, or about 98° W. *From that point we desire that every effort be used to endeavor to penetrate to the southward and westward, in a course as direct towards Bering's Strait as the position and extent of the ice, or the existence of land at present unknown, may admit.*

'SECTION 6. We direct you to *this particular part* of the Polar Sea as affording the *best prospect* of accomplishing *the Passage* to the Pacific, . . . but should your progress in the direction before ordered be arrested by ice of a permanent appearance, and if when passing the mouth of the strait between Devon and Cornwallis islands, you had observed that it was open and clear of ice, we desire that you will duly consider . . . whether *that channel* might not offer a more practicable outlet . . . and a more ready access to the open sea.'

FRANKLIN sailed 26th May 1845 with 129 souls, provisioned to July 1848, and from Whale Fish Islands sent his last letter to the Admiralty, 12th July 1845, but it contained no definite information. The *Erebus* and *Terror* were last seen by a whaling captain, DANNETT,

26th July 1845, moored to an iceberg, 74° 40′ N., 66° 13′ W., waiting for an opening in the middle ice so as to cross to Lancaster Sound. Thus FRANKLIN and his expedition vanished forever from the sight of civilized man.

Lancaster Sound must have been practically ice-free, but the expedition found progress toward Cape Walker prevented by unfavorable ice-conditions in the eastern part of Barrow Strait. In this contingency FRANKLIN followed the alternative allowed him, and ascended Wellington Channel, — then only known through PARRY, 1819, as an opening to the northward.

Marked success attended their efforts, as FRANKLIN attained 77° N. latitude, a higher point than any successor reached except BELCHER. The extent of his necessarily extensive discoveries in the polar sea, to the north of Wellington Channel is unknown, but FRANKLIN returned southward by the west of Cornwallis Land, thus proving it to be an island, — a fact that escaped later explorers.

The ice to the west forbidding farther progress that season, the *Erebus* and *Terror* went into winter quarters at Beechey Island, 74° 42′ N., 91° 32′ W. An observatory and workshop were built on shore, sledge journeys were made to the east and north, and with returning summer there were even attempts at a garden. Three men, BRAINE, HARTNELL, and TORRINGTON, here died. As soon as the opening ice afforded an opportunity of advance the squadron left Beechey Island, hurriedly some think, as no record was left.

From Cape Walker FRANKLIN'S course is uncertain. BROWN, in *North-West Passage*, thinks that he passed through McClintock Strait, west of Prince of Wales Land, while MCCLINTOCK believes, with great reason, that, entering Peel Sound, FRANKLIN sailed by the most direct route down Franklin Strait, along the west coasts of North Somerset and Boothia. In either event, the squadron reached Victoria Strait, where both ships were beset, 12th September 1846, in the open sea, twelve miles north of King William Land, in 70° 05′ N., 98° 23′ W. Although the Northwest Passage was almost completed, — they were within 90 miles of the known sea of America, — the situation necessarily caused FRANKLIN much anxiety, the more so as the autumn passed without either vessel being released.

The winter, 1846–47, passed without especial fatalities, and with opening spring, 1847, FRANKLIN turned his attention to adjacent and unknown shores. We may well believe he would first explore King William Land, between Point Victory, of ROSS, 29th May 1830, and Cape Herschell, of SIMPSON, 25th August 1839. While FRANKLIN was planning this sledge journey from his ice-beset ships, off the *west* coast of Boothia, RAE was exploring the Boothian peninsula, where he reached, 18th April 1847, a point less than 150 miles from FRANKLIN. One of FRANKLIN'S sledge parties, under Lieutenant GORE, left the *Erebus*, 24th May 1847, and in June deposited at Point Victory, in a cairn of ROSS, 1831, the following paper, which, found by MCCLINTOCK'S party in 1859, is the only record of the expedition from 1845 to 1847:

'28th of May, 1847.

H. M. ships *Erebus* and *Terror* wintered in the ice, in latitude 70° 05′ N., longitude 98° 23′ W.

'Having wintered in 1846–47 [An error. The correct dates should be 1845–46. See dates at top and bottom of record] at Beechey Island, in latitude 74° 43′ 28″ N., longitude 91° 39′ 15″ W., after having ascended Wellington Channel to latitude 77°, and returned by the west side of Cornwallis Island.

'Sir JOHN FRANKLIN commanding the expedition. All

well. Party consisting of 2 officers and 6 men left the ships on Monday, 24th May, 1847.

GM. GORE, *Lieut.*
CHAS. F. DES VOEUX, *Mate.*'

On 11th June 1847, FRANKLIN ended his Northwest search by quiet death on the ice-beset *Erebus*. His passing is beautifully chronicled by Lord TENNYSON on the memorial tablet in Westminster Abbey: —

'Not here: the white North has thy bones; and thou,
Heroic sailor-soul,
Art passing on thine happier voyage now
Toward no earthly pole.'

By FRANKLIN'S death the command devolved on CROZIER. What steps were taken to explore the adjacent coasts, to open communication with the natives, to exploit the resources of the country, or to reconnoitre the line of retreat is unknown; but preparations for the abandonment of the vessels necessarily began when the summer of 1847 passed without the ice breaking up. Inroads were made in their numbers the third year by disease, doubtless fostered by enforced inactivity and a reduction of rations, for they were provisioned only to July 1848. Before the retreat 24 men died, among them a large number of officers.

By 1848 the ships had drifted 19 miles southwest of their place of besetment, and CROZIER abandoning them started for Back River. Landing at Point Victory, the record of GORE, 1847, was brought to their camp, and on its margin this record was written: —

'April 25, 1848. — H. M. Ships *Terror* and *Erebus* were deserted on the 22nd of April, 5 leagues N. N. W. of this, having been beset since 12th September, 1846. The officers and crews, consisting of 105 souls, under the command of Capt. F. R. M. CROZIER, landed in latitude 69° 37′ 42″ N., longitude 98° 41′ W. This paper was found by Lt. IRVING, under the cairn supposed to have been built by Sir JAMES ROSS, in 1831, 4 miles to the northward, where it had been deposited by the late Commander GORE, in June, 1847. Sir JAMES ROSS' pillar has not, however, been found; and the paper has been transferred to this position, which is that in which Sir JAMES ROSS' pillar was erected. Sir JOHN FRANKLIN died on the 11th June, 1847, and the total loss by death in the Expedition has been to this date 9 officers and 15 men.

F. R. M. CROZIER, *Captain and Senior Officer.*
JAMES FITZJAMES, *Captain H. M. S. Erebus.*

And start on to-morrow, 26th, for BACK'S Fish River.'

How a party of 105 men could almost vanish in the limited region around King William Land has seemed remarkable. Three winters had necessarily affected their health; among CROZIER'S party must have been many in the last stages of disease, and many others must have been tainted by incipient scurvy. Actuated by the determination that usually characterizes Arctic commanders, no doubt exists that CROZIER held together his men, sick and well, as long as the faintest hope remained, and there is not the slightest indication that the miserable cry of *Sauve qui peut!* was ever raised.

The journey to Back River could not be made by CROZIER in less than 250 miles. Their movements are not definitely known, despite MCCLINTOCK'S thorough and magnificent search in 1859 (Chapter XI). His researches, supplemented by Eskimo narratives, have been fully confirmed by the later labors of HALL, 1869, and the remarkable search from the summer encampment of SCHWATKA and GILDER on King William Land, 1879.

Struggling along the west coast of King William Land, each day must have made their fate more evident. Provisions probably failed near the south end of this barren land, as undeniably a party returned to the ships, where a body was found by the Eskimo before the vessels sank or stranded. A small party of Eskimo saw and camped with some of the retreating party, but fearing their common safety would be compromised by remaining, the natives stole away, leaving the white men to their fate. By graves and skeletons the line of retreat is traced from Point Victory to Todd Island, south of King William Land, and there are reasons to believe that some reached Point Ogle and others Montreal Island; but with one and all it was death by disease, or worse — by starvation.

That they met death with courage, loyalty, and solidarity is indisputable. The old Eskimo woman paid the highest tribute possible to her ideal, that of physical merit, when she said to McClintock, 'They fell down and died as they walked,' which was verified by the position of a skeleton found by McClintock himself. Faithful to the last these heroic men, as Sir John Richardson beautifully says, 'forged the last link of the Northwest Passage with their lives.'

McClintock: *Fate of Sir John Franklin; Voyage of Fox* (5th ed. London 1881); Nourse: *Hall's Second Arctic Expedition* (Washington 1879); Gilder: *Schwatka's Search for the Franklin Records* (New York 1880).

CHAPTER X

THE FRANKLIN SEARCH BY LAND

THE first uneasiness regarding the fate of the Franklin party was excited by communications from Sir JOHN ROSS to the Admiralty during the winter of 1846–47, but nothing, beyond offered rewards to whalers to examine Lancaster Sound, was done until 1848, when the Admiralty was fully aroused. In co-operation with the sea-search (Chapter XI) RICHARDSON, already familiar with the ground, was to descend the Mackenzie, examine the shore to the Coppermine, coast the west and south shores of Wollaston Land, and search the straits to the east and west, so as to cross detached parties of Sir JAMES ROSS, who was to operate by ship from Barrow Strait. RICHARDSON secured the invaluable services of Dr JOHN RAE, and in a boat voyage of great difficulty thoroughly examined the coast between the Mackenzie and Cape Kendall. Here their boats were cached, 3d September 1848, owing to the very early winter which obliged the party to reach overland their winter quarters, Fort Confidence, 66° 54′ N., 118° 49′ W.

In 1849, a single boat remained, and RICHARDSON having the fullest confidence in the judgment, experience, and prudence of D^r^ RAE, sent him northward to examine the shores of Wollaston and Victoria lands. In thus acting RICHARDSON nobly sacrificed all personal considerations, and in selecting RAE, said that his zeal, ability, personal activity, and skill as a hunter fitted him peculiarly for the enterprise. Leaving Fort Confidence 9th June 1849, RAE reached twelve days later Kendall River, where supplies had been sent in April. The solid ice of the Coppermine broke so slowly that it was 13th July before RAE could enter the sea, and twelve days were consumed in reaching his cached boats of 1848. Eskimo along the coast were most friendly, but they knew nothing of FRANKLIN's party. RAE spent 24 days at Cape Krusenstern in an unavailing attempt to cross Dolphin Strait to Wollaston Land, but gales and ice alike forbade.

Renewing operations in 1851, under the auspices of the Hudson Bay Company, RAE left Fort Confidence, where he had wintered, 25th April, and with two men reached on foot the Polar Sea, at the Coppermine. A journey to the westward brought them, 9th May, to 68° 38′ N., 110° 02′ W., whence they turned west to Douglas Island. Crossing Dolphin Strait, RAE was the first white man to visit Wollaston Land; he traced its western coast and, 22d May 1851, reached Cape Baring, 70° 00′ N., 117° 17′ W., of Prince Albert Sound. Recrossing the strait to Cape Krusenstern, Kendall River was reached, 10th June. This foot journey of 1,100 miles, made by Dr RAE in 33 days, on a daily ration of two pounds per man, is one of the most remarkable on record. At Kendall River he met, as previously arranged, a boat-party from Fort Confidence, but they were unable to leave the Coppermine until 8th June. Passing through Dease Strait, RAE reached Cape Colburn 1st August, and commenced his examination of the east coast of Victoria Land, where his discoveries began at Cape Princess Royal, 6th August 1851. Detained by packed ice, RAE left his boat and, travelling on foot, attained his farthest northing, 70° 03′ N., 101° 25′ W., 12th August. RAE's boat was now in the waters, and within 50 miles of the spot where the *Erebus* and *Terror* were abandoned, three years and four months earlier. FRANKLIN and RAE thus made by boat and ship the nearest approach to the Northwest Passage by *sea;* the next nearest was by two ships, the *Hecla*, PARRY 1819, and the *Enterprise*, COLLINSON, 1851. RAE unfortunately failed in his efforts to cross Victoria Strait to King William Land, which was in sight, else he might have recovered the Franklin records, or discovered the stranded ship and learned the fate of the party four years earlier than he did. On his return he found at Parker Bay the butt of a flagstaff, with tack and line bearing the government mark, doubtless from the Franklin squadron.

The next land expedition, also under Dr RAE, produced the first evidence of the fate of FRANKLIN'S men. Leaving Chesterfield inlet by boat, 10th August 1853, RAE reached Repulse Bay four days later, killing *en route* a walrus that furnished lamp-oil for the winter. RAE had three months' provisions, but whether he could winter yet depended on successful hunting or Eskimo aid. He pitched his tent with gloomy prospects, game scarce, no natives, and no traces of late visits. RAE determined to remain until the last moment, and endeavor to obtain enough fuel and food to render wintering possible. On 1st September, fixed for their return, they had fuel for 14 weeks, but had not been very successful in the chase. While prospects were not encouraging the party to a man volunteered to risk the winter. Fortunately game became abundant, and by 1st October they were safely provided, having killed 109 deer, one musk-ox, 106 ptarmigan, one seal, and caught 190 salmon, — 49 deer and the musk-ox being due to RAE'S personal skill.

The winter over, a cache for spring journeys was laid down at Lady Pelly Bay, and 31st March 1854 RAE started to explore the west coast of Boothia. On 20th April, in 68° 29′ N., 90° 19′ W., he met a young Eskimo, who gave him the first information obtained by civilized man of the fate of the FRANKLIN expedition.

In the spring of 1850, about 40 white men were seen dragging a boat southward along the west shore of King William Land. They bought a seal from Eskimo hunters, whom they told that their ship had been crushed by ice, and that they were going to a land where they could shoot reindeer. Later that spring, before the ice broke up, the bodies of some 30 men were found on the continent, and five on an island a day's march to the northward. This pointed to the Eskimo encampment of Back River and Montreal Island as the places, though possibly they referred to Starvation Cove, of SCHWATKA, or Tod Island, of HALL, both near the mouth of Back River. The natives reinforced their statements by producing silver with the Franklin crest, which, with other articles, left no doubt that their story was substantially correct, and that the Franklin expedition had perished. Numerous relics of the Franklin squadron were obtained from the natives and brought back to Churchill that autumn.

RAE was compelled to hunt and explore on foot, without dog-sledge or native assistance of any kind. Under these disadvantages it is not to be considered surprising that he did not explore all of west Boothia; it is rather a matter of congratulation that his geographic discoveries were so extended. To the east he reached Castor and Pollux River, thus connecting with SIMPSON, 20th August 1839, and discovered Murchison River, 27th April 1854. To the north along the west coast of Boothia, RAE's farthest was on 6th May 1854, 68° 58′ N., 94° 22′ W., beyond which he sent an Indian six miles, very near Cape Porter, ROSS, 1831. The summer game and Eskimo aid would have rendered the complete exploration of the west Boothia coast an easy matter; but RAE, returning to give to the world his information of the fate of FRANKLIN, reached Fort York in August 1854.

In 1855, JAMES ANDERSON, of the Hudson Bay Company, descended Back River in three canoes, meeting Eskimo at various points. On 30th July, at the lower rapids, Eskimo were found with many Franklin relics. The natives said they came from a boat belonging to white men, who had died of starvation. On Montreal Island, 1st August, there were Eskimo caches, in which were many additional relics, but ten days' search around the mouth of Back River gave no additional information. Unfortunately the expedition had no interpreter, was inadequately equipped, and could not cross to King William Land owing to the frailty of their boats.

On the return of ANDERSON with indefinite information, confirmatory of other indefinite information, the Admiralty considered the fate of FRANKLIN determined, and awarded to D^r RAE and his companions £10,000, the offered reward to any one setting at rest the fate of FRANKLIN and his companions, which had been done indirectly. Thus ended the exertions of the British Admiralty to determine the exact fate and extend succor to the unfortunate members of its official expedition for the Northwest Passage. It remained for a wife's devotion, at private expense, to ascertain that which the government of Great Britain acknowledged as its duty, but which its officialism was unable to accomplish.

Nine years after ANDERSON, came an American, C. F. HALL, who had spent two years, 1860–62, with natives near Frobisher Bay, where he had found relics of FROBISHER'S three voyages. Determined to discover the Franklin record, HALL landed at Depot Island, Hudson

Bay, 20th August 1864, with two natives, a whale-boat, tent, and moderate amount of supplies. Waiting unsuccessfully a year for Eskimo aid, he proceeded to the old winter-quarters of RAE, Fort Hope, Repulse Bay, and there wintered, 1865–66; and in the spring got as far as Cape Weyton, 68° N., 89° W., beyond which point his Eskimo refused to go. Here, however, he met other natives who had visited the deserted ships, and had seen FRANKLIN. From these Eskimo HALL obtained considerable silver, bearing the crest of FRANKLIN and of other officers. Then, unable to reach King William Land, the determined HALL visited, in February 1867, Igloolik, winter-quarters of PARRY, 1822. In 1868, following up the west side of Melville Peninsula, HALL did some geographic work of interest by completing the short gap between RAE'S farthest, 1846, and PARRY'S farthest in Fury Strait, 1825; thus filling in the last bit of the north coast-line of the continent of America.

Wintering again at Fort Hope, 1868–69, HALL at last succeeded in securing the Eskimo aid for which he had patiently waited five years. Having accumulated supplies, he started, March 1869, with ten Eskimo and dog-sledges. Rae Peninsula was crossed to Committee Bay, and via Boothia Isthmus, of ROSS, 1831, they reached James Ross Strait, within 60 miles of King William Land. With great reluctance the natives consented to go west of Pelly Bay, but at Simpson Island a successful musk-ox hunt put them in humor to proceed. At Point Ackland, eastern end of Ross Strait, they fell in with an Eskimo who guided them to Tod Island, South of King William Land, where a human thigh bone was found. On 12th May 1869, HALL put foot on the mainland, but the only tangible result of three days' search was a human skeleton. In returning he met other natives who had personal or traditional knowledge of the Franklin disaster; and their reports were to the effect that CROZIER with 40 men had dragged two sledges down the west coast of King William Land the last of July 1848, and near Cape Herschel met four Eskimo families, whom he told he was going to Repulse Bay. The natives stole away, and the party died of starvation. One ship drifted southwest to O'Reilly Island, 68° 30′ N., 99° W., which if true completes the entire Northwest passage by ships. Information was obtained pointing to the direct fate of 79 of the 105 retreating men, leaving 26 to reach and perish on the American coast, probably at Montreal Island.

The final land-search was made by Lieutenant F. SCHWATKA, U. S. Army, and W. H. GILDER, who wintered 1878–79 among the natives near Chesterfield Inlet, Hudson Bay. Enlisting the Eskimo in his scheme, which to them was an extended hunting party, SCHWATKA started, April 1879, with four whites, 14 Eskimo, food for one month, and abundant ammunition. Travelling overland to Back River he met natives, one of whom remembered BACK, 1834, and another had visited FRANKLIN'S ships. Near Montreal Island and Point Richardson, 31st May 1879, SCHWATKA found a hundred natives, who assisted in exploring exhaustively the continental coast-line to Point Seaforth, south of King William Land. The search was fruitless, and SCHWATKA was convinced that the story of a cairn and records, on which the expedition had been based, was groundless. However, he was not the man to lose a great opportunity; he determined on the daring plan of crossing Simpson Strait to King William Land, and camping for the summer to search thoroughly the ground traversed by the retreating party. SCHWATKA'S journey to the island had entailed some 450 miles of travel and occupied 70 days, and the crossing to King

William Land necessarily meant a five months' stay, until the ice of autumn made it possible to return to the mainland. To appreciate the hazard of this journey it should be remembered that the land visited, substantially unknown, was believed to be devoid of game.

Crossing to King William Land, 10th June, SCHWATKA and GILDER established a summer camp as a base of operations. With four Eskimo and dog-sledges SCHWATKA and his three white companions carefully searched the whole island, giving three months to the work. Four despoiled graves and six unburied skeletons were found, and the remains of the boat discovered by MCCLINTOCK in 1859 (Chapter XI). The tenting-place in Terror Bay, of which the Eskimo had said much both to HALL and SCHWATKA, was not found, presumably having been obliterated by the encroachments of the sea. Many relics of the unfortunate men were collected, brought back, and presented to the English government. The record deposited by MCCLINTOCK, 3d June 1859, was also found, though the cairn had been destroyed by Eskimo. From native accounts it appears that four men survived to 1849, for their footsteps in the spring snow were seen by Eskimo on King William Land, where they were probably hunting reindeer.

In the autumn SCHWATKA and his companions thoroughly examined the point to the westward of Point Richardson, Starvation Cove, where the natives told HALL there were the remains of a boat and of some 30 men. From native accounts there are good reasons to believe that with this boat were the records brought from the ship, and also the magnetic needle used in making observations at the magnetic pole, that FRANKLIN crossed in reaching the place of his final besetment. Five miles inland was found the skeleton of a man, probably one of the last survivors, who perished searching for food.

The expedition of SCHWATKA and GILDER made no important additions to geographic knowledge, for this country had been traversed by many parties. Had it been otherwise their contributions must have been extensive, for the journey is one of the most remarkable in the annals of Arctic sledging. They were absent from their original base of supplies a year less ten days, during which they travelled 2,819 geographic miles. It is difficult to say which most to admire, the daring of the plan or the skill that wrought its success.

In (YOUNG'S) *Cruise of the Pandora*, 1876 (Chapter xi), MACGAHAN thus describes near the scene of death the fate of the 'last man' of the Franklin expedition: —

'One sees this man, after the death of his last remaining companions, all alone in that terrible world, gazing round him in mute despair, the sole living thing in that dark, frozen universe. The sky is sombre, the earth whitened with a glittering whiteness that chills the heart. His clothing is covered with frozen snow, his face lean and haggard, his beard a cluster of icicles. The setting sun looks back to see the last wretched victim die. He meets her sinister gaze with a steady eye, as though bidding her defiance. For a few minutes they glare at each other, then the curtain is drawn and all is dark.'

RICHARDSON: *Boat Journey through Rupert Land* (London 1851); NOURSE: *Hall's Second Arctic Expedition* (Washington 1879); GILDER: *Schwatka's Search for the Franklin Records* (New York 1880). For RAE, ANDERSON, and HOOPER, see *Arctic Blue Books*, Chapter xviii.

CHAPTER XI

THE FRANKLIN SEARCH BY SEA

THE Sea Search falls naturally into two divisions, from the Atlantic and from the Pacific. The first Pacific squadron was under Captain T. E. L. MOORE, *Plover*, with Captain H. KELLETT, *Herald*. The *Plover* was too late to pass Bering Strait, but the *Herald* spent a month of the autumn of 1848 in Kotzebue Sound, where later the two ships and SHEDDON's yacht *Nancy Dawson* rendezvoused in July 1849. SHEDDON was the first to round Point Barrow by ship, accompanying so far Lieutenant W. J. PULLEN, who, with three boats, examined the coast eastward to the mouth of the Mackenzie, where he arrived 5th September. Ascending the river he wintered at Fort Simpson, met Dr RAE there, and under new orders unsuccessfully endeavored in 1850 farther coast explorations east and north.

Faithfully exploring the waters north of Bering Strait the *Herald* reached, 29th July 1849, 72° 51′ N., 163° 48′ W., discovered one of the isolated (Herald) islands of the Siberian Sea, and possibly saw Wrangell Island.

The *Plover* wintered in Kotzebue Sound, 1849–50, and reached Dease Inlet in 1850. MAGUIRE, taking her with a new crew, in 1851, explored by boat that autumn to Return Reef (FRANKLIN, 1827). The winters 1851–52, 1853–54, were passed at Point Barrow, the intervening quarters being at Point Clarence. The seasons at Barrow were utilized by Dr JOHN SIMPSON in studying the natives, and these observations, supplemented by others during his five years in this region, resulted in the first, and in some respects the most important, memoir ever published on the western Eskimo.

A second squadron was sent to operate *via* Bering Strait, the *Enterprise*, Captain RICHARD COLLINSON, commanding, and the *Investigator*, Captain ROBERT M'CLURE. Leaving England 11th January 1850, the ships met again only at Magellan Strait, for M'CLURE reaching first Bering Strait, 31st July, declined to wait 48 hours for COLLINSON, and six days later avoided communication with the *Plover*. The fleet instructions read: 'We caution you against suffering the two vessels under your orders to separate, except in the event of accident or unavoidable necessity. [You are] in no way to hazard the safety of ships and the lives intrusted to your care, by your being shut up in a position which might render a failure of provisions possible. The object of the expedition is to obtain intelligence and to render assistance to Sir JOHN FRANKLIN and his companions, and not for the purpose of geographical or scientific research.' Ignoring these orders, without prearranged rendezvous or definite coöperative plan, M'CLURE rounding Point Barrow, 5th August 1850, grounded his ship, lost 3,300 pounds of meat, and pushed his ship so far into the main pack that he barely escaped besetment. Communicating with natives at Cape Bathurst, he turned north and landed, 7th September, on a new (Banks) land, and following a (Prince of Wales) strait to the east was beset in the middle of it on 11th September. Drifting a few miles south, the *Investigator*, in 72° 52′ N., 117° 03′ W., experienced for nine months the horrors of the pack, which often threatened to destroy the ship and obliged M'CLURE to land supplies on Princess Royal Islands.

M'CLURE, starting 25th October, with a sledge party following the strait to its northern end, reached the north-east extremity of Banks Land (PARRY, 1819), and overlooked the water-ways navigated by that explorer. This journey established the *then* earliest known existence of continuous water communication north of America, although we *now* know that an earlier and shorter route was discovered by FRANKLIN, 1846–47, in attaining SIMPSON'S farthest (Chapter IX).

In July 1851, M'CLURE unavailingly endeavored to sail northward into Barrow Strait; reaching, 14th August, 73° 14′ N., he was obliged to pass Banks Land by the south. The navigation along the west coast was daring and dangerous in the extreme. OSBORN says: 'The coast became as abrupt and precipitous as a wall; the water was very deep . . . 15 fathoms when touching the cliffs on one hand or the lofty ice on the other. The pack was of fearful description; it drew 40 feet of water, and rose in rolling hills, some 100 feet from base to summit . . . . Nothing in the long tale of Arctic research is finer than the cool and resolute way in which this gallant band fought their way around this frightful coast.'

Especial dangers need not be enumerated, but it seems as if on several occasions the *Investigator* was almost saved by Providence. Reaching the extreme northwest point of Banks Land she was beset, 20th August to 11th September, when she fortunately escaped, and two days later was in the waters of Barrow Strait.

Attempting night navigation, 23d September, M'CLURE'S ship grounded at the entrance of a bay (Mercy). Here in 74° N., 118° W., the winter of 1851–52 was passed comfortably, game being abundant. M'CLURE visited Melville Island and found, 29th April 1852, at the winter quarters of PARRY (1819–20), a cairn with MCCLINTOCK'S record, dated 6th June 1851, stating that AUSTIN's expedition had wintered between Cornwallis and Griffith islands; from which M'CLURE correctly assumed that AUSTIN returned to England later in 1851. Other sledge journeys were made in 1852 as follows: Lieutenant CRESSWELL, 32 days, 170 miles out, explored the north and northwest coast of Banks Land and later followed Wollaston Land south to look for the *Enterprise*. Lieutenant HASWELL, in 47 days, reached 70° 45′ N., 114° W., 14th May; eight days later RAE (page 134) reached a point in this inlet only 40 miles distant. HASWELL met natives living on Wollaston Land, and M'CLURE communicating learned from them of the continuity of Wollaston and Victoria lands. Lieutenant WYNNIATT reached, 26th May, Prince Patrick Land, within 60 miles of OSBORN's farthest (page 135.)

The summer of 1852 brought no chance of escape, and the following winter a reduction of rations was necessary, as game was far less plentiful. The returning sun of February 1853 found the party in a most precarious condition. M'CLURE decided to retreat, one party to go east to Cape Spencer, 550 miles distant, whence they might meet whalers of Baffin Bay. The other party going south was to take the small cache at Princess Royal Island, and reach by boat the north coast of America and the Hudson Bay posts. The parties were told off 3d March, and to put them in condition for the march, 15th April, were given full rations. This last test of human endurance was not exacted of the party, which was so reduced in health that their able surgeon, Dr ARMSTRONG, says the journey would have proved fatal; in any event, three men died the first half of April. On 6th April 1853, as they were making a grave, a sledge-party suddenly appeared, from AUSTIN's squadron wintering to the eastward, under command of Lieutenant BEDFORD PIM, who, says ARMSTRONG, had most providentially reached the *Investigator*, after 'a most severe and harassing journey of 28 days, being then the earliest polar traveller on record.' Later under orders from Sir EDWARD BELCHER, commanding the eastern squadron, the *Investigator* was abandoned, and its crew crossing by sledge the ice of Barrow Strait was the first and last party that ever made the Northwest Passage.

M'CLURE's voyage was geographically a grand success, but otherwise it must be classed as a failure. Disregarding the spirit of the official instructions he lost his ship, and but for the almost miraculous appearance of PIM would have sacrificed his crew. Had M'CLURE awaited COLLINSON, the chances of discovering the Franklin party, *in extremis* or all dead, would have been greatly enhanced, for stronger sledging parties could have reached King William Land. Of M'CLURE's voyage Admiral Sir HENRY RICHARD says: 'But for his chief's [COLLINSON] unsuspicious and trusting nature [he] would never [have] had the opportunity of making himself famous.' The noble character of COLLINSON was later displayed in his defence of M'CLURE, whose actions and success materially militated against COLLINSON's future.

Returning to the flag ship *Enterprise*, COLLINSON reached Cape Lisburne, 13th August 1850, two weeks behind M'CLURE, and met with the great polar pack eight days later in 72° N., 153° W. As COLLINSON had not met the *Herald*, and so was ignorant that the slow *Investigator* had passed eastward, he decided to run no risks, but to follow orders. Tracing the edge of the solid pack he reached, 28th August, 73° 23′ N., 164° W., having passed over the reported land of the *Plover*, Captain MOORE, 29th July 1849.

Returning south for the winter, COLLINSON entered again the Arctic Ocean in 1851 and passed from Point Barrow eastward, 31st July. Entering Prince of Wales Strait he examined M'CLURE's cache on Princess Royal Islands, and later reached the entrance of Barrow Strait, where, 31st August, he was in 73° 30′ N., 114° 35′ W., some distance beyond the farthest attained by M'CLURE in 1850. Being within 57 miles of the farthest western point reached by the *Hecla* (PARRY, 1819), it is the nearest approach of ships in the Northwest Passage. Turning southward COLLINSON followed M'CLURE so closely (he had missed him coming out of his winter quarters only by ten days) that when COLLINSON sailed up the east coast of Banks Land to his farthest, 72° 52′ N., 125° W., 7th September 1851, he was less than a hundred miles behind, for the *Investigator* was then beset in 74° 25′ N., 122° W. Thinking from lack of cairns that M'CLURE had not followed this coast, COLLINSON wisely turned back and saved his ship.

The *Enterprise* wintered in Walker Bay, 71° 36′ N., 117° 41′ W., where she was frozen in, 21st October. Some forty Eskimo, summer inhabitants near Walker Bay, moved south in November and returned in May.

Parties were put in the field to discover the *Investigator*, and to search for FRANKLIN. COLLINSON and seven men, tracing the northwest coast of Prince Albert Land, reached in a journey of 52 days and 537 miles Glenig Bay, 13th May 1852. Lieutenant JAGO and eight men camped, 10th May, at the head of Prince Albert Sound, 70° 43′ N., 110° 45′ W. Lieutenant PARKES visited Melville Island to search Winter harbor (PARRY, 1819), but east of Cape Providence he saw sledge tracks, thought he heard dogs, and fearing natives, as he was unarmed, turned back with two men. It is the only time in all the Franklin search that an officer avoided other parties: he was on dangerous service in any event. and had he gone on he would have learned that the *Investigator* was in Mercy Bay and that McClintock of Austin's squadron had searched that coast in June 1851.

Breaking out of Walker Bay, August 1852, Collinson surveyed Prince Albert Sound, and then turned his prow toward Dolphin Strait, that was to lead him of all government expeditions nearest to the fateful remains of the Franklin expedition. Following the continental coast of America, and passing through the isle-bestrewed Coronation Gulf, he reached the east end of Dease Strait and went into winter quarters, at Cambridge Bay, 69° N., 105° W., 28th September 1852.

Resuming sledge travel in 1853, Collinson in a journey of 49 days searched the southeast coast of Victoria Land, — picking up Rae's record of 13th August 1851 (page 134), — and attaining his farthest, 10th May, at Gateshead Island, 70° 26′ N., 100° 47′ W. Here he looked east across the frozen strait, where Franklin's ship had sunk, to King William Land, unconscious that there lay the unburied skeletons of the men he sought. Rough ice and weak sledge parties forbade crossing, and underweight coal put on board at Plymouth left no fuel for another winter.

Collinson from two sources had traces of Franklin. The natives of Cambridge Bay had a steam-engine rod, and an article marked with the Queen's broad arrow, but his interpreter being with M'Clure, no one was able to gather information as to the source whence the article came. Again in July 1853, Collinson found on Finlayson Island part of a ship's door or hatchway, which, as his boatswain, formerly of the *Erebus*, did not recognize, Collinson ascribed to the *Victoria* abandoned *east* of Boothia (page 96).

On 11th August 1853, Collinson left Cambridge Bay and retraced his way through the difficult and intricate passages to the westward. Unable to pass Camden Bay, owing to ice, he went into winter quarters, 29th September, at Flaxman Island, some 200 miles west of Point Barrow, in 70° N., 145° W. He was able to leave 12th July 1854, and England was reached 5th May 1855, after an absence of five years and four months.

The voyage of COLLINSON is one of the most remarkable and successful on record. With a *sailing* ship he navigated not only the Arctic Sea forward and back through 120 (64 one way) degrees of longitude, a feat only excelled by the *steamer Vega*, but he also sailed the *Enterprise* more than ten degrees of longitude through the narrow straits along the northern shores of continental America, which never before nor since have been navigated, save by small boats and with excessive difficulty. Of all government naval expeditions searching for FRANKLIN he came nearest the goal. COLLINSON'S modest journal is characterized by Admiral RICHARDS, one of the few living men fully competent to pass on the merits of Arctic work, as 'a record of patience, endurance and unflagging perseverance, under difficulties which have perhaps never been surpassed.'

The most persistent and extensive exertions for the relief of FRANKLIN were those made by sea from the Atlantic. The first expedition entrusted to Sir JAMES C. ROSS, an able officer of Arctic service, did not add to his reputation. Crossing the middle ice of Baffin Bay, ROSS reached with the *Investigator* and *Enterprise* open water in 75° 05′ N., 68° W., 20th August 1848. Searching the coast between Pond and Possession Bays, and visiting such portions of the north shore of Barrow Strait as the ice would permit, he ran into Port Leopold, 11th September, where he was shut in for the winter. In May 1849, ROSS and Lieutenant MCCLINTOCK explored by sledge all of Prince Regent Inlet and the northern gulf of Boothia, except 160 miles between Fury beach and Lord Mayor Bay. Meanwhile in 1849, the *North Star*, under Master SAUNDERS, was sent out with provisions to refit Ross, but she failed to cross the middle ice of Baffin Bay and wintered in Wostenholme Sound. In 1850 SAUNDERS was again unable to reach Port Leopold, and landing his supplies near Wollaston Island reached England in September, where he unexpectedly found Ross. The summer of 1849 gone and the ice unbroken, Ross had erected a house, filled it with supplies, and at the earliest practicable date, 28th August, cut his way out of Port Leopold. Bad fortune pursued him; most unfortunately his ships were speedily beset, and despite every effort could not be extricated till 25th September, off Pond Bay, whence he returned to England. 'Altogether,' as BROWN says, 'this was a most unfortunate expedition.'

The spring of 1851 was full of bustle in connection with the Franklin Sea Search. The Admiralty organized two expeditions, British private generosity sent forth a third, a spirit of sympathy impelled American citizens and its government to unite in equipping a fourth, and the wifely devotion of Lady FRANKLIN put yet another ship in commission. Unfortunately they all tended in one direction, Lancaster Sound, and to especial efforts on the shores of Wellington Sound, hundreds of miles from the proper place. One action of the British government was unparalleled, for it put in the same field two independent expeditions. The whaling captain, WILLIAM PENNY, sailed in the *Lady Franklin*, with A. STEWART in the *Sophia*, while the larger expedition was intrusted to Captain HORATIO AUSTIN, R. N., with Captain OMMANEY, Lieutenant OSBORN, and Lieutenant CATOR in command respectively of the *Assistance*, *Intrepid*, and *Pioneer*. A private expedition was under Sir JOHN ROSS, in the *Felix* with Commander G. PHILIPS. The United States was represented by Lieutenant E. J. DE HAVEN, in the *Advance*, and M[r] S. P. GRIFFIN in the *Rescue*, with the heroic KANE as fleet surgeon. Lady FRANKLIN'S vessel was the *Prince Albert*, Commander C. FORSYTH and M[r] W. P. SNOW.

These ten vessels reached the southeast entrance of Wellington Channel nearly together. The first signs — distinct traces of Europeans — were found by Captain OMMANEY at Cape Riley and Beechey Island, 23d August 1851. The impression that the encampment was of the Franklin expedition was changed to certainty when PENNY, 27th August, found at Beechey Island three graves of men of the *Erebus* and *Terror*, who had died between January and April, 1846. There were other signs of the wintering, but an exhaustive search resulted in no other information beyond that given by scattered articles and the lonely graves.

FORSYTH at once carried the news to England in the *Prince Albert*, which ship it may be here said sailed again, May 1851, under Captain WILLIAM KENNEDY, with a French volunteer, Lieutenant J. R. BELLOT. She wintered at Batty Bay, after hazardous experiences during which KENNEDY and four men were separated from her six weeks, taking refuge at Somerset house (ROSS, 1832–33). KENNEDY and BELLOT made a very long sledge journey, 1,100 miles in 97 days, of great importance. They discovered that Brentford Bay, of ROSS, was a (Bellot) strait, reached 21st April 1852, 100° W., visited Cape Walker *via* Franklin Strait, and travelled entirely around North Somerset. Altogether the KENNEDY search was one of the best conducted and most promising of all, relative to the end in view.

DE HAVEN decided to return home, but strong gales

and severe cold beset and froze-in his ships. Beset in the middle of Wellington Channel, the American squadron drifted to the north, attaining 75° 25′ N., 93° 31′ W., where they discovered Murdaugh Island, and, beyond North Devon, an extensive land, which they called Grinnell; the credit of this discovery was ungraciously contested. The drift changed to the south in October 1850, and later to the east. Month after month, in darkness and solitude they moved slowly through Wellington Channel and Lancaster Sound, travelling 1,050 miles with the floes to which they were indissolubly bound for over eight months. Life on the ships was almost unendurable with its anxieties, monotony, and privations. Disruption of the pack threatened almost daily, with prospects of a winter on the naked floe under conditions of disease, darkness, and biting cold. July released the shattered vessels, but DE HAVEN felt obliged to abandon the search, and reached the United States, 30th September 1851.

The *Lady Franklin*, *Sophia*, and *Felix* wintered in Assistance Bay at the south end of Cornwallis Land. In the spring of 1851 PHILLIPS of ROSS's party crossed Cornwallis Island. PENNY in a journey to Hamilton Island found, 18th May 1851, his progress stopped by open water, which reached as far north in Wellington Channel as he could see, and led him to believe that the missing squadron was in that quarter. STEWART and SUTHERLAND of his squadron explored the east side of Wellington Channel between Capes Eden and Belcher, and GOODSIR following the west side explored the north coast of Cornwallis Land, reaching its northwest extremity, and just stopped short of discovering that the land was an island.

The great sledging work, however, was that of AUSTIN's squadron. The preparations and successful accomplishment of these journeys indicate forethought and executive ability of no slight order on the part of AUSTIN; and in turn the determination, endurance and energy of the officers and men were every way worthy of the Royal Navy. The squadron was frozen in at Griffith Island, in September 1850, when autumnal journeys to lay out depots for spring travel at once began.

In 1851 AUSTIN intrusted to Captain OMMANEY the principal search in the vicinity of Cape Walker, the place to which his instructions carried FRANKLIN. OMMANEY, with Lieutenants OSBORN, BROWNE, and MECHAM, discovered and outlined the north half of Prince of Wales Island, and in his report properly remarks that the coast search was 'exactly in the route where Sir JOHN FRANKLIN was instructed to seek a passage to the American continent.' His comments on the northern coast, which was shoal and ice-lined with grounded floes, indicate its unnavigability. There was no trace of FRANKLIN either at Cape Walker or on any point of Prince of Wales Land, which justifies the opinion that he never landed thereon, but followed its east coast down Peel Sound. In his *Northwest Passage*, BROWNE remarks: 'Previous to the journey of this excellent officer (OMMANEY), no attempt had been made to reach Cape Walker, or to follow the Franklin expedition in the direction ordered in Section 5 of his instructions, and yet six years had passed, and it was known that the expedition was victualled for only three.'

Other sledging divisions equally distinguished themselves by their field-work, the journeys of MCCLINTOCK and BRADFORD being particularly creditable. The energy displayed, and geographic work done, by AUSTIN'S expedition are illustrated in the following table: —

COMMANDERS	MEN	DAYS OUT	MILES TRAVELLED	NEW COAST DISCOVERED	FARTHEST POINT REACHED				DATE 1851
Capt. E. OMMANEY .	6	60	480	205	72°	44′ N.	100°	42′ W.	24th May
Lt. S. OSBORN . . .	7	58	506	70	72	18	103	25	23d "
Lt. W. H. BROWNE .	6	44	375	150	72	49	96	40	13th "
Lt. R. D. ALDRICH .	7	62	550	70	76	16	104	30	17th "
Lt. McCLINTOCK . .	6	80	760	40	74	38	114	20	28th "
Surg. BRADFORD . .	6	80	669	135	76	23	106	15	22d "

AUSTIN concluded that FRANKLIN did not proceed to the south or west of Wellington Channel, and deemed it unnecessary to prosecute farther search in these directions, while PENNY believed that nothing more could be done to the north: consequently both decided to return to England that autumn, 1851. Their differences and estrangement led to a parliamentary investigation, which was not detrimental to either expedition.

The Arctic committee, however, recommended another effort *via* Barrow Strait, and with strange fatuity said: 'We consider no farther exploration . . . to the southward of Cape Walker necessary, and therefore propose that all the energy of the expedition be directed toward the examination of the upper portion of Wellington Strait.' The expedition consisted of Sir EDWARD BELCHER commanding; *Assistance*, Commander GEORGE H. RICHARDS; *Resolute*, Captain HENRY KELLETT; *Pioneer*, Lieutenant SHERARD OSBORN; *Intrepid*, Commander F. L. McCLINTOCK, and transport *North Star*, Lieutenant PULLEN.

The squadron reached Wellington Channel, 14th August, whence Dr M'CORMICK in a hazardous boat-journey examined the lower part of the strait. BELCHER sent KELLETT and McCLINTOCK to the west, and with OSBORN ascended Wellington Strait to an isle-covered (Northumberland) sound, where he wintered in 76° 52′ N.,

97° w. With RICHARDS, his most efficient coadjutor, and OSBORN he discovered that autumn Exmouth and Cornwall Islands, the latter in 77° 34′ N., 97° W. RICHARDS and OSBORN took the field with sledges, 10th April 1853. They examined thoroughly Cornwallis, Bathurst and Melville Islands, making many new discoveries in their remarkable journey of 860 miles, in 94 days. BELCHER starting northeast reached a (Belcher) channel, 76° 31′ N., 90° W., 20th May 1853, and returned through Arthur Strait, by the east side of Grinnell Land, thus proving it to be an island. In a second trip he reached, 10th June, an (Buckingham) island, 77° 10′ N.

The *Resolute*, KELLETT, and *Intrepid*, MCCLINTOCK, reached Melville Island, September 1852, but being unable to enter Winter Harbor returned to Dealy Island, where they were frozen in, 10th September. During the autumn and spring, 1852–53, the following sledging was done, in days of service and miles of travel, by this division of the squadron: MCCLINTOCK, 145 days, 1,661 miles; MECHAM, 117 and 1,375; R. ROCHE, 79 and 1,039; G. S. NARES, 94 and 980; Dr DOMVILLE, 77 and 739, and DE BRAY, 62 days and 642 miles. Altogether the travel amounted to 8,558 miles.

The most notable spring journeys were: —

COMMANDER	TOTAL PARTY	DAYS OUT	MILES TRAVELLED	FARTHEST REACHED		DATE 1853
Com. MCCLINTOCK . . .	9	105	1401	77° 23′ N.	118° 20′ W.	17th June
Lieut. G. F. MECHAM . .	8	94	1163	77 06	120 30	29th May
Mate G. S. NARES . . .	8	69	980	75 32	119 30	3d "
Lieut. B. V. HAMILTON . .	8	68	974	76 38	104 50	26th "
Lieut. BEDFORD PIM . . .	8	62	635	74 06	117 56	6th "
Dr DOMVILLE	8	76	739	74 06	116 56	21st "

MCCLINTOCK, already famous as the greatest of Arctic

sledgemen, surpassed himself by a journey remarkable for its duration, distance, and success, while MECHAM was scarcely second.

By all means the most important results were those arising from the journey of Lieutenant BEDFORD PIM, of the *Resolute*. The autumnal journey of MECHAM to establish depots on Melville Island led him to Winter Harbor, where he discovered the record left by M'CLURE the preceding spring. Fearing that the *Investigator* might have been detained at Mercy Bay, KELLETT at the earliest moment despatched PIM, who took the field 10th March. His sledge broke down before the journey was half accomplished, but fortunately PIM appreciated the gravity of the situation. Leaving DOMVILLE to follow as best he could, PIM with two men and a dog-sledge pushed on and reached the *Investigator*, by the earliest extended spring journey on record, — 160 miles in 28 days. Had PIM turned back it would have been fatal to some of the crew of the *Investigator*, for he reached the ship only nine days prior to the date fixed by M'CLURE for her abandonment. It is safe to say that PIM's determination, judgment, and exertions averted another Franklin disaster, and instead enabled the crew of the *Investigator* to complete the Northwest Passage.

The summer of 1853 was very backward, Barrow Strait failed to break up in July, and BELCHER had to face the contingency of another winter, with his position weakened by the disabled crew of the *Investigator* draining his supplies. Under these circumstances he ordered the abandonment of the *Resolute* and *Enterprise*, which were fast-bound in the unbroken ice far to the west of Lancaster Sound. But the ice to the eastward still held firm around his own ships, and it was not until 5th August that the first ship, *North Star*, was in open water. Determining to avoid at any cost a third winter in the ice, BELCHER ordered the abandonment of the *Assistance* and the *Pioneer*. The crew of the five ships were assembled on the remaining vessel, *North Star*, and on 25th August she turned homeward. Barely had she started when two vessels, the *Phœnix* and *Talbot*, were sighted, which strangely enough bore orders practically identical with the line of action of BELCHER. The Lords of the Admiralty and their representatives in Arctic waters tacitly declared, despite the fact that FRANKLIN's fate was yet unsettled, that this was the last of Arctic voyages.

The sledging feats of MCCLINTOCK, MECHAM, and others, and the accomplishment of the Northwest Passage by M'CLURE, excited enthusiastic and deserved praise in Great Britain, but exultations over these heroic deeds of British seamen were not unmingled with feelings of shame and indignation that the outcome of these expeditions was the abandonment of five ships of the Royal Navy, and that such action was taken without determining the fate of FRANKLIN. M'CLURE and his men, sent out under double pay for the relief of FRANKLIN and not for geographic exploration, claimed the reward for the Northwest Passage and were granted £10,000; but the skeletons of FRANKLIN's men were left unfound and unburied.

It may appropriately be added that BELCHER's action in avoiding another winter and in abandoning farther search for FRANKLIN to the north or west of Lancaster Sound is amply justified by present knowledge. The strong feeling in England against BELCHER was later emphasized by the remarkable voyage of the *Resolute*. This ship, abandoned by BELCHER's orders, 15th May 1854, in 74° 41′ N., 101° W., withstood the ice in 1855, and drifting a thousand miles through Barrow Strait, Lancaster Sound, and Baffin Bay, was discovered north of Cape Dyer, in

67° N., and brought safely to port by Captain J. M. BUDDINGTON, an American whaler. Congress appropriated $40,000 for the purchase of the *Resolute*, and refitting her presented her to the Queen and people of Great Britain as a token of good-will on the part of the American people. Ships'-stores, flags, officers' libraries, and so on, had been preserved, and were restored to their original position, so that to her old officers she appeared to be exactly as she was when abandoned in the Arctic Seas.

But if the British government was content to leave the Arctic mystery unsolved, the wifely devotion of Lady FRANKLIN viewed the matter differently. Of the £35,000 spent by private parties for the relief of FRANKLIN, far the larger part came from her personal fortune. Her prescient judgment and undaunted perseverance outlined the correct plan and furnished the means; and had KENNEDY been able to go south from Bellot Strait the question would have been solved years earlier. The discoveries of Dr RAE (page 135) had determined that the expedition had perished, but Lady FRANKLIN insisted on knowing the whole tale. As the British government declined to continue the search, she decided to expend all her available means for a final effort.

Most fortunately she secured the services of Captain LEOPOLD MCCLINTOCK, who had signally distinguished himself under ROSS (1848–49), AUSTIN (1850–51), and KELLETT (1853–54). MCCLINTOCK sailed in the steam yacht *Fox*, 1st July 1857, with two able and energetic officers, Captain ALLEN YOUNG and Lieutenant W. R. HOBSON, R. N. Beset in the middle of Melville Bay, the *Fox* drifted eight months, 1,200 miles to the south. The hardships of an Arctic winter are sufficiently depressing at the best, but when experienced under conditions where the fate of ship and crew hang for months in the balance, and at a time when deferred action seems to offer certain failure, it is difficult to understand how McClintock and his officers endured them. They did more, however, for when they escaped, after a winter in the drifting pack, their renewal of the search is unparalleled in the history of Arctic service.

Refitting in the Greenland ports, McClintock reached Beechey Island, and there erected a monument to the Franklin expedition. Following Peel Sound south — in the very track of Franklin — and meeting solid ice, they turned on their tracks and anchored in Port Leopold, 19th August 1858. Examining the stores here left by J. C. Ross in 1849, McClintock sought the uncertain passage to the west (Brentford Bay of John Ross, 1829) thought by Kennedy to be a (Bellot) strait. McClintock found it a nominal passage to the west, for his most strenuous efforts failed to force the *Fox* through, although he made five determined attempts. He was eventually obliged to go into winter quarters at its northeast extremity, in Port Kennedy, 72° N., 94° W.

Advance depots for spring travel were laid out that autumn, and on 17th February 1859, McClintock started, in a temperature of 60° to 70° below freezing, to communicate with the Boothians, whom he found twelve days later when encamped at the north magnetic pole. There were 45 natives well provided with relics of a (Franklin) party of 'white people [who] starved upon an [Montreal] island where there is a [Back] river.' McClintock returned to the *Fox* 14th March, having travelled 360 miles, practically completed the coast line of continental America, and added 110 miles of new land to the charts.

On 2d April, McClintock and Hobson started on their final journey, each with a man-sledge and a dog-sledge.

Meeting the natives, from whom McClintock bought many relics, they said: Two ships had been seen near King William Land; one sank, and the other was forced on the shore by ice and broken up; the ships were destroyed in the autumn, and all the white people, taking boats, went away to the large river, and the following winter their bones were found there. Cape Victoria was reached, 28th April 1859, whence Hobson was sent direct to Cape Felix, King William Land, to search the west coast for the stranded ship and records; and if there unsuccessful to carry out the original plan of completing the discovery and search of Victoria Land, between the extremes reached by Collinson and Winnyatt (page 146).

McClintock followed the coast southward and encamped, 1st May, near Port Parry. Six days later he fell in with 40 Eskimo, who sold him silver plate and other relics. They informed him that the ship had disappeared, that there were many books long since destroyed by weather, and that the wreck had last been visited in the winter of 1857–58. An old woman who had visited the wreck said that many of the white men perished at Back River; some were buried and others not. Following the east coast of King William Land, McClintock crossed to Point Ogle, and 15th May reached Montreal Island. A thorough search resulted in but few traces of Europeans, among which, however, were a few remnants of boat-fittings. The coast and bays of Simpson Strait were carefully examined without result, and the party crossed to King William Land, 24th May. The next night McClintock came upon a bleached human skeleton lying on its face, in a position which indicated that the man, suffering from hunger and exhaustion, had fallen and died in that posture, thus confirming the truth of the words of the old Eskimo woman, who said: 'They fell down and died as they walked.'

The most important relics were in Erebus Bay, northeast of Cape Crozier, where HOBSON, preceding MCCLINTOCK, had discovered a boat on a sledge, pointing northward as though it had been abandoned while the party were travelling in that direction. In the boat were two human skeletons, and around it large quantities of clothing and all kinds of odds-and-ends, of great weight and little use; the only provisions were 40 pounds of chocolate and a little tea. Near Cape Crozier, MCCLINTOCK learned from records of HOBSON that *this* coast had been thoroughly explored without finding the slightest trace of records or natives, and, what was of the greatest importance, that he had found at Point Victoria, on the northwest coast of King William Land, a record, the first and last direct information that has ever come from the Franklin party.

This record (pages 129–131) briefly sets forth their winter at Beechey Island, their important geographic discoveries in Wellington Channel, the besetment of their vessels near King William Land, the death of FRANKLIN, eight other officers, and 15 men, the abandonment of the ships, 25th April 1848, after two years' besetment, and the efforts of CROZIER, with 105 souls, to reach Back River.

HOBSON came first upon traces of FRANKLIN'S expedition west of Cape Felix, where he found a large cairn, three tents, a small English ensign, and many less important articles. So thorough was his search of the coast, says MCCLINTOCK, that 'coming over the same ground after him, I could not discover any traces that had escaped him.'

MCCLINTOCK and HOBSON reached the *Fox*, 14th and 19th June, respectively, the former having visited Montreal Island, completing the exploration and circuit of King William Land, while HOBSON had found the Franklin record.

YOUNG was still absent. With a four-man-sledge and a dog-team he commenced explorations 7th April, and finding a channel between Prince of Wales Land and Victoria Land, sent his men back, and for 40 days travelled with one man and a dog-sledge. Driven back to the *Fox* by the great exposure and the fatigue, which seriously impaired his health, he entered the field again after three days' rest, despite the protest of Dr WALKER. YOUNG was in the field 78 days under most discouraging circumstances. He crossed Franklin Strait to Prince of Wales Land, traced its shores to its southern termination at Cape Swinburne, and attempted to cross McClintock Channel, but the ice was too rough to render the journey practicable with the means and time at his disposal. He completed the exploration of this coast beyond OSBORN'S farthest to nearly 73° N., and in addition explored both shores of Franklin Strait between the *Fox* and ROSS'S farthest in 1849, and BROWN'S in 1851. In all, YOUNG explored 380 miles of new coast, which, with 420 miles discovered by MCCLINTOCK and HOBSON, made a magnificent contribution of 800 geographic miles of new shore-line.

It may here be added that YOUNG, in 1875, made a gallant attempt to sail through Peel and Franklin straits, pass east of King William Land, and reach Bering Strait. Unfortunately, he was forced back by an impassable ice-barrier in Peel Strait, 72° 14′ N. (See YOUNG: *The Two Voyages of the Pandora:* London, 1879.)

Breaking out of Port Kennedy, 4th August 1859, MCCLINTOCK was obliged, owing to the death of his engineer, to stand at the engine 24 consecutive hours. The return journey was difficult, but without serious mishap, and the *Fox* reached Portsmouth, 24th September 1859, where definite information of the fate of FRANKLIN'S expedition for the first time reached the civilized world.

The news of FRANKLIN'S Northwest Passage and accompanying geographic triumphs, of his besetment and death with 23 others, of CROZIER'S fateful retreat and perishing men, swept instantly over the world, awakening sympathy for the dead, and winning plaudits for the bravery and skill of the living who had wrested from the silent North the story of Sir JOHN FRANKLIN and his crew.

See *British Arctic Blue Books,* Chapter xviii; SEEMAN, end Chapter vi; OSBORN and ARMSTRONG, end Chapter viii; SNOW: *Voyage Prince Albert,* 1850 (London 1851); KENNEDY: *Second Voyage Prince Albert* (London 1853); SUTHERLAND: *Voyage Lady Franklin and Sophia,* 2 v. (London 1853); KANE: *First Grinnell Arctic Expedition* (New York 1854); BELCHER: *Last of Arctic Voyages,* 2 v. (London 1855); McDOUGALL: *Voyage H. M. S. Resolute,* 1852–54 (London 1857); BROWN: *Northwest Passage* (London, 2d edition, 1860); McCLINTOCK: *Fate of Sir John Franklin; The Voyage of the Fox* (London, 5th edition, 1881); COLLINSON: *Journal of H. M. S. Enterprise* (London 1889).

CHAPTER XII

NORTH-POLAR VOYAGES

CONTRARY to the general impression, Arctic voyages for reaching the north geographic pole have been the exception rather than the rule. ROBERT THORNE successfully urged on HENRY the Eighth a renewal of the search for a short northern route to China by sailing across the North Pole; but of the resulting expedition in 1527, which consisted of two ships, only its failure is definitely known.

Eighty years later, HENRY HUDSON's first recorded voyage, 1st May 1607, was to discover a passage by the North Pole to Japan and China. The heavy ice of the Spitzbergen Sea stopped him in 80° 23′ N., off the east coast of Greenland, 13th July. Although the North Pole was not crossed, far-reaching results followed, which are set forth in the account of Spitzbergen. JONAS POOLE's voyages of 1610–11 turned to fishing ventures without passing to the north of Spitzbergen.

The next voyage of note is that of Captain J. C. PHIPPS, afterward Lord MULGRAVE, who sailed, 4th June 1773, in the *Racehorse*, with Captain S. LUTWIDGE in the *Carcass*. Following the west coast, ice was fallen in with at the northwest extremity of Spitzbergen, and the edge of the ice-barrier was traced some ten degrees to the west until 9th July, in 80° 36′ N., when PHIPPS found farther progress impossible. Eventually he was compelled to abandon the voyage, having reached 80° 48′ N., 20° E., a higher latitude than any of his predecessors. Especial interest attaches to this expedition as the immortal NELSON, then a lad of fifteen, served as PHIPPS's coxswain, and by a foolhardy adventure with a bear is said to have barely escaped death, as he persisted in attacking the animal under most dangerous circumstances.

In the revival of Arctic explorations in 1817, Great Britain decided to send two of its four polar vessels to reach the Pacific by crossing the North Pole. Captain D. BUCHAN commanded, in the *Dorothea*, with Lieutenant JOHN FRANKLIN, just recovered from his wound of New Orleans, in the *Trent*. Sailing 25th April 1818, and rendezvousing in June at Magdalena Bay, Spitzbergen, BUCHAN was able to sail no farther north than 80° 37′ N., where his ships were beset. A severe storm on 30th July freed the ships, but the *Dorothea* was so badly crushed by frequent nips of the ice-pack that she was in danger of foundering. The *Trent* convoyed her to a Spitzbergen port, and in October they returned to England.

Whalers are not concerned in crossing the Pole, but doubtless some of the Dutch skippers, hundreds of whom passed annually for a century the 80th parallel, found the open water and a single day's favorable wind that would have carried them beyond 83° or 84° N. latitude. Exact data are wanting as to their exploits, and the story of sailing across the Pole may well be discredited. However, SCORESBY, the famous British whaler, passed far beyond any other authenticated northing, when he reached 81° 30′ N., 19° E., 24th May 1806.

Holding fast to the Spitzbergen route for the Pole, unquestionably the most promising of *navigable* routes, PARRY attempted in 1827 to reach the Pole. Profiting by his experiences in four previous Arctic voyages, PARRY most carefully equipped his expedition, which sailed in the *Hecla*. The ship was safe in Trurenberg Bay 20th June, and here the only change of programme occurred, as PARRY decided to leave behind the tame reindeer he had brought as draught animals. The very next day PARRY left with two boats (fitted with steel-shod runners so as to serve as sledges), 28 souls, and provisions for 71 days. The final departure from land was made 23d June, at Little Table, one of the Seven Islands.

Ice was fallen in with immediately, and progress was exceedingly slow and fatiguing, owing to unfavorable conditions which continued to the end. Fog and rain were astonishingly frequent, the ice-floes were loose, small, and exceedingly rugged, the water-pools of small extent and not connecting, and the ice itself was largely in the shape of irregular, needle-like crystals placed vertically, nearly close together and pointed at both ends. As summer advanced the needles became movable, rendering it extremely fatiguing to walk over them, besides cutting both boots and feet. Changes from floe to water and back occurred several times each march, which commenced about six P.M. and lasted till early morning. Sometimes the boats had to be moved up or down almost perpendicular slopes, and frequently the same road had to be travelled five times, as the whole load could not be advanced at the same time.

The daily northing averaged about five miles at first, but it gradually dropped, as indicated by the following daily latitudes: 5th July, 81° 45′ N.; 10th, 82° 03′; 12th, 82° 11′; 13th, 82° 17′; 17th, 82° 31′; 20th, 82° 36′; 21st, 82° 39′; 22d, 82° 43′; 24th, estimated 82° 45′; and 26th July, 82° 42′.

On 20th July, however, PARRY made a discovery that destroyed his hopes of making a very high latitude. In three days his dead reckoning showed that he had made twelve miles to the north, but his observations gave less than five miles. Sixteen hours of effort the following day enabled them to advance between ten and eleven miles, but the observations gave not quite four miles. They had drifted to the south nearly seven miles during their rest and sleep of eight hours, and the southerly drift since 16th July had averaged four miles per day.

The extreme point reached, 82° 45′ N., 20° E., was 172 miles from the *Hecla*, by a free water-route of 100 miles and of 192 miles over floes and water-pools, raised by doubling travel to a total of 580 miles, about the distance from the *Hecla* to the Pole; the entire journey covered 61 days and 970 miles.

The expedition failed to obtain the £1,000 offered if they reached 83° N., but they secured for England a new record of the highest latitude, 82° 45′ N., which remained for 48 years the farthest north. PARRY'S success was received with the greatest enthusiasm in Great Britain, and distinctions at home and from abroad were freely conferred on him. FRANKLIN returned from Arctic America at the same time, and both the distinguished polar travellers were knighted by the Crown and honored by degrees from Oxford.

The voyages of KANE, 1853, and HAYES, 1859, were not open attempts to reach the North Pole, though incidentally the leaders looked in that direction. Their expeditions are treated in connection with Smith Sound.

The next North-polar expedition, that of Sweden in the *Sofia*, was organized by NORDENSKIÖLD, who accompanied it as the scientific chief, the naval commander being Captain (Count) F. W. VON OTTER. The plan was to reach the northernmost accessible port of Spitzbergen in early autumn, and leaving on a favorable occasion attain by ship as high a northing as possible. Smeerenberg Bay was the place of rendezvous, whence various scientific journeys were made. In early September the *Sofia* made an unsuccessful attempt to reach the Seven Islands. Leaving Smeerenberg again, 16th September, von Otter found a heavy ice-pack in the vicinity of Seven Islands, but by doubling and pushing he reached, 19th September 1868, 81° 42′ N., 17° 30′ W., the highest latitude attained before or since by ship in the Eastern hemisphere. Elsewhere it has been exceeded only by Hall, *Polaris*, 1870; Nares, *Alert* and *Discovery*, 1875; and Greely, *Proteus*, 1881.

A third attempt failed, owing to the *Sofia* being injured by collision with a floe, which necessitated her return that autumn to Sweden. In addition to the then unprecedently high latitude by ship, Nordenskiöld made such valuable collections that Heer said he had achieved more, and broadened more the horizon of knowledge, than if he had merely reached the North Pole.

Although not next in chronological order, it is best here to mention the Swedish polar expedition of 1872, organized by Nordenskiöld with the expectation of reaching the Pole by reindeer-sledging from a ship in North Spitzbergen as his base of operations. The *Polhem* was put under Lieutenant Palander, Swedish navy, and made its winter-quarters in Mussel Bay. Elsewhere (Chapter V) have been set forth the unfortunate conditions under which the supply ships were frozen in, with many Norwegian walrus hunters. Though his reindeer were lost and his party weakened by their generous treatment of the men thrust upon them, yet Nordenskiöld and his companions did not abandon their spring work. Three sledges left Mussel Bay 24th April 1873; but two breaking down, and other accidents occurring, Nordenskiöld did not reach Phipps Island until 17th May, then with a single sledge. The ice to the north was so exceedingly rough that it was evident no progress could be made, and NORDENSKIÖLD, wisely abandoning the effort to the north, returned to the *Polhem* by the way of the inland ice of North-East Land (Chapter V).

The first German North-polar expedition was fitted out in 1868 through the exertions of Dr PETERMAN, who sent Captain K. KOLDEWEY in the *Germania*. Unable to reach East Greenland, which was to be the line of approach, KOLDEWEY sailed to Spitzbergen, reached 81° 05′ N. off the north coast, and sailing down Henlopen Strait sighted Wiche Land, and reached home that autumn.

The second German expedition fitted out immediately, with KOLDEWEY commanding in the *Germania*, and Captain HEGEMAN in the *Hansa*. The *Germania* reached the rendezvous, Sabine Island, East Greenland, and its fortunes there are elsewhere considered (Chapter XVII). The *Hansa*, separating by misunderstanding from her consort, was frozen in the pack, 5th September, and drifting south with the ice was crushed, 19th October. A house was built on the floe, where the party barely escaped death on several occasions, when violent storms broke up the main ice. The shipwrecked men, who had drifted over 600 miles along, and in sight of, the barren coast of Greenland, on 7th May, after 200 days' drift, took to their boats in 61° 12′ N., and rounding Cape Farewell reached Friedrichstal, 13th June 1871.

The United States in 1870 sent Mr C. F. HALL on a North-polar expedition in the *Polaris*, and his discoveries are considered in connection with Smith Sound. It may be here said that HALL attained in the Polar Ocean 82° 11′ N., then the highest north ever reached by ship, and now exceeded only by NARES, 1876, in the *Alert*.

The British polar expedition of 1875 following Smith Sound is treated elsewhere (Chapter XIV), except the effort to reach the North Pole, which was made from Floeberg Beach, the *Alert's* winter quarters, 82° 24′ N. The command was entrusted to a gallant officer of great ability, Commander A. H. MARKHAM. Supported as far as Cape Henry, 82° 55′ N., MARKHAM took the frozen polar sea, 10th April 1876, with 2 sledges and 17 men.

The dread of the open polar sea overshadowed their prospects by loading the sledgemen with two boats, which so weighed them down that they advanced with divided loads, and the 73 miles made good from the ship necessitated 276 miles of travel. One boat was soon abandoned, but the other was dragged to 83° 20′ N., 64° W., the farthest north to that time, and only once exceeded. This point was reached 12th May 1876, by exertions that taxed human endurance to the utmost, owing to the indescribably rugged conditions of the ice-floes. Disease disabled the party long before it turned back, when not less than five of the 17 men were on the sledge, disabled by scurvy. That such a northing was made at all was due to the persistent energy of MARKHAM and the heroic determination of his men. During the return journey matters went from bad to worse; the second boat was abandoned, the disabled increased in number, and the whole party would have perished but for the extraordinary march of twenty-four hours made by Lieutenant A. C. PARR to the *Alert*, whence aid was obtained. One man died *en route*, and eleven others of the original 17 were carried to the ship on sledges. No party ever strove harder, nor more deserved success, than did that of MARKHAM, and that their success was marred by disaster arose from causes beyond control,—incipient scurvy and over-loading.

The Peary success of 1891 and his failure of 1894 are too intimately associated with Greenland (Chapter XVII) for treatment here. Undiscouraged by two failures to reach a very high latitude, this energetic and tenacious explorer yet (1895) continues his efforts.

In 1894 WALTER WELLMAN, an American, endeavored to reach the Pole by sledge and boat, using a ship at Spitzbergen as his base. Most improved scientific devices were adopted to reduce weights and secure success, but unfortunately the *Ragnvald Tarl* was crushed by floes, 28th May, at Walden Island. Although the ice was exceedingly rough, undeterred by adverse conditions WELLMAN pluckily continued north by sledge after the wreck, but was soon obliged to abandon the attempt 6 miles east of Platen Islands, near the 81st parallel.

Dr NANSEN in 1893 initiated a novel and most dangerous plan. Ignoring the accepted canons of ice-navigation, of avoiding besetment and of following the protected lee of land-masses, he avowed his intention of putting his ship into the great ice-pack northeast of the Kara Sea; thence he expects to be carried by the ice-drift across the Pole. Provisioned for five years, NANSEN hopes to solve the problem in three years. His ship, the *Fram*, passed safely into Kara Sea, but to September, 1895, no word has been had of his farther movements.

In July, 1894, Mr. F. JACKSON sailed from England, in the *Windward* for a North-polar voyage by the most promising of all *land* routes. Landing his dogs, horses, houses, equipments, and supplies at a suitable point on Franz Josef Land, he was to send back his ship that autumn. JACKSON was to prosecute his explorations to the northward at his leisure, and have his vessel return at an appointed time. As the *Windward* has not been heard from (September, 1895) since the end of August, 1894,—then in 75° 45′ N., 44° E.,—it is evident that JACKSON decided on the safer course of retaining his ship, or that besetment or other danger has disarranged his plans.

Supplementary to these efforts, the following tables are of interest as marking progressions toward the Pole: —

RECORDS OF THE HIGHEST NORTH MADE SINCE 1587 (*in the Eastern and Western Hemispheres, by land and by sea*)

EASTERN HEMISPHERE

COMMANDER	DATE	N. LAT.	LONG.	LOCALITY
WILLIAM BARENTS . .	14th July 1594	77° 20′	62° E.	Near Cape Nassau, N Z.
RYP & HEEMSKERCK . (BARENTS' 3d Voyage)	19th June 1596	79 49	12 E.	N. Spitzbergen
HENRY HUDSON . . .	13th July 1607	80 23	10 E.	Spitzbergen Sea
J. C. PHIPPS	27th July 1773	80 48	20 E.	" "
WILLIAM SCORESBY .	24th May 1806	81 30	19 E.	" "
W. E. PARRY	23d July 1827	82 45	20 E.	" "
NORDENSKIÖLD & OTTER	19th Sept. 1868	81 42	18 E.	Spitzbergen Sea, highest by *ship*
WEYPRECHT & PAYER	12th April 1874	82 05	60 E.	Franz Josef Land, by PAYER, highest land

WESTERN HEMISPHERE

COMMANDER	DATE	N. LAT.	LONG.	LOCALITY
JOHN DAVIS	30th June 1587	72° 12′	56° W.	W. Greenland
HENRY HUDSON . . .	20th June 1607	73	20 W.	Off E. Greenland
WILLIAM BAFFIN . .	4th July 1616	77 45	72 W.	Smith Sound
E. A. INGLEFIELD . .	27th Aug. 1852	78 21	74 W.	Smith Sound
E. K. KANE	24th June 1854	80 10	67 W.	Cape Constitution, Greenland, by MORTON
C. F. HALL	30th Aug. 1870	82 11	61 W.	Frozen Sea
C. F. HALL	30th June 1871	82 07	59 W.	Greenland, by Sergeant F. MEYER, Signal Corps, U. S. A.
G. S. NARES	25th Sept. 1875	82 48	65 W.	Grinnell Land, by ALDRICH
G. S. NARES	12th May 1876	83 20	65 W.	Frozen Sea, by A. H. MARKHAM
A. W. GREELY . . .	13th May 1882	83 24	41 W.	New Land, north of Greenland, by LOCKWOOD & BRAINARD

Doubtless the name of some whaler should follow that of BAFFIN in the above list, but the inexactitude of most high latitudes reported by whalers is well known. Possibly the reported northing of LAMBERT, 78½° N., in 1670 on the East Greenland coast, may have exceeded INGLEFIELD'S exact latitude of 78° 21′ N.

Sweden holds the ship record in the old world, but PARRY beat it by boat. It will be noted that England held the honors of the farthest north through HUDSON, 1607; PHIPPS, 1773; PARRY, 1827; and NARES, by ALDRICH, 1875, and by MARKHAM, 1876. This record, unbroken for 275 years, passed to the United States through the efforts of the International Polar expedition under Lieutenant GREELY, which by LOCKWOOD and BRAINARD reached 83° 24′, the most northerly point, whether on sea or land, ever attained by man (Chapter xvi).

PHIPPS: *Voyage towards North Pole* (London 1774); PARRY: *Attempt to reach the North Pole in boats* (London 1828); BEECHEY: *Voyage towards North Pole* (London 1843); MARKHAM, A. H.: *Great Frozen Sea* (London 1878); LESLIE: *Nordenskiöld's Arctic Voyages* (London 1879); KOLDEWEY *et al.*: *Second German North-Polar Expedition* (London 1874); SMITH, *Tyson's Arctic Experiences* (N. Y. 1874); DAVIS: *Hall's (Polaris) North-Polar Expedition* (Washington 1876).

CHAPTER XIII

THE ISLANDS OF THE SIBERIAN OCEAN

BETWEEN Bering Strait and the Kara Sea the shallow Siberian Ocean is dotted with scattered islands; the most important of these, commonly known as Liachof or New Siberian Islands, are situated to the northeast of the Lena Delta. The most southerly island, Liachof, had been seen by traders coasting from the Lena to the Indigirka, but it was never visited until 1770, and then by a Russian trader, LIACHOF.

His journey was inspired by the sight of an immense herd of deer coming south from the ice-clad sea. Following their tracks by dog-sledge in April 1770, he reached in a day an island (Liachof) and the following day another (Maloi), 60 miles from the continent. Obtaining exclusive right of exploiting the land he had explored, LIACHOF visited the islands by boat in 1773, wintered on Liachof, and discovered a third (Kotelnoi) island, on which were tusks of the mammoth. The discovery of mammoth bones in 1750 by LERCHON on the Siberian tundra initiated a profitable industry which now extended to these islands.

The scientific and material importance of these deposits led the Russian government to send CHVOINOF to survey these islands, and obtain information regarding their remarkable natural conditions. His reports showed that save a few granitic hills the soil of Liachof was a mixture of ice, sand, and ivory, the last being remains of the mammoth, fossil ox, rhinoceros, and other animals. Among other remarkable conditions was that of ice overgrown by moss to a considerable depth. From a high mountain on Kotelnoi CHVOINOF saw in May a mountainous land to the north, doubtless De Long Islands. Driftwood was abundant, ivory plentiful, foxes and other valuable fur-bearing animals very numerous.

At the beginning of this century other islands were discovered by hunters, — Stolbowoi and Fadejef, or Thaddeus, in 1805 by SAMKIF, Nova Sibir the following year by SIROVATSKOF, while BJEFLKOF added in 1808 the small islet of his name.

To the fortunate complications regarding hunting-rights are due the descriptions of HEDENSTRÖM, so important to scientists. Sent by and at the expense of Chancellor NICHOLAS ROMANZOF to survey the islands, HEDENSTRÖM left Ustjansk 7th March 1809, with SANNIKOF and KOSHEVIN as assistants, and travelling by sledge reached Thaddeus *via* Liachof. He continued on to New Siberia Island, while the two others surveyed Thaddeus and Kotelnoi. In 1809, with SANNIKOF, he again visited Nova Sibir, which was reached with 29 sledges, 13th March. On this island was found an axe made of a mammoth tusk, and on Thaddeus a Jukahir sledge and skinning knife, which indicate that these natives had once lived on these islands, at a remote period before iron was obtainable from the Russians. On the south coast of Nova Sibir were discovered the remarkable Wood Hills, 200 feet high, consisting of alternate strata of sandstone and bituminous tree-remains. The wood was friable, black, glossy, and at times like fossilized charcoal. HEDENSTRÖM verified the almost incredible statement of LIACHOF that the whole soil of Liachof Island appears to consist of mammoth bones. In any

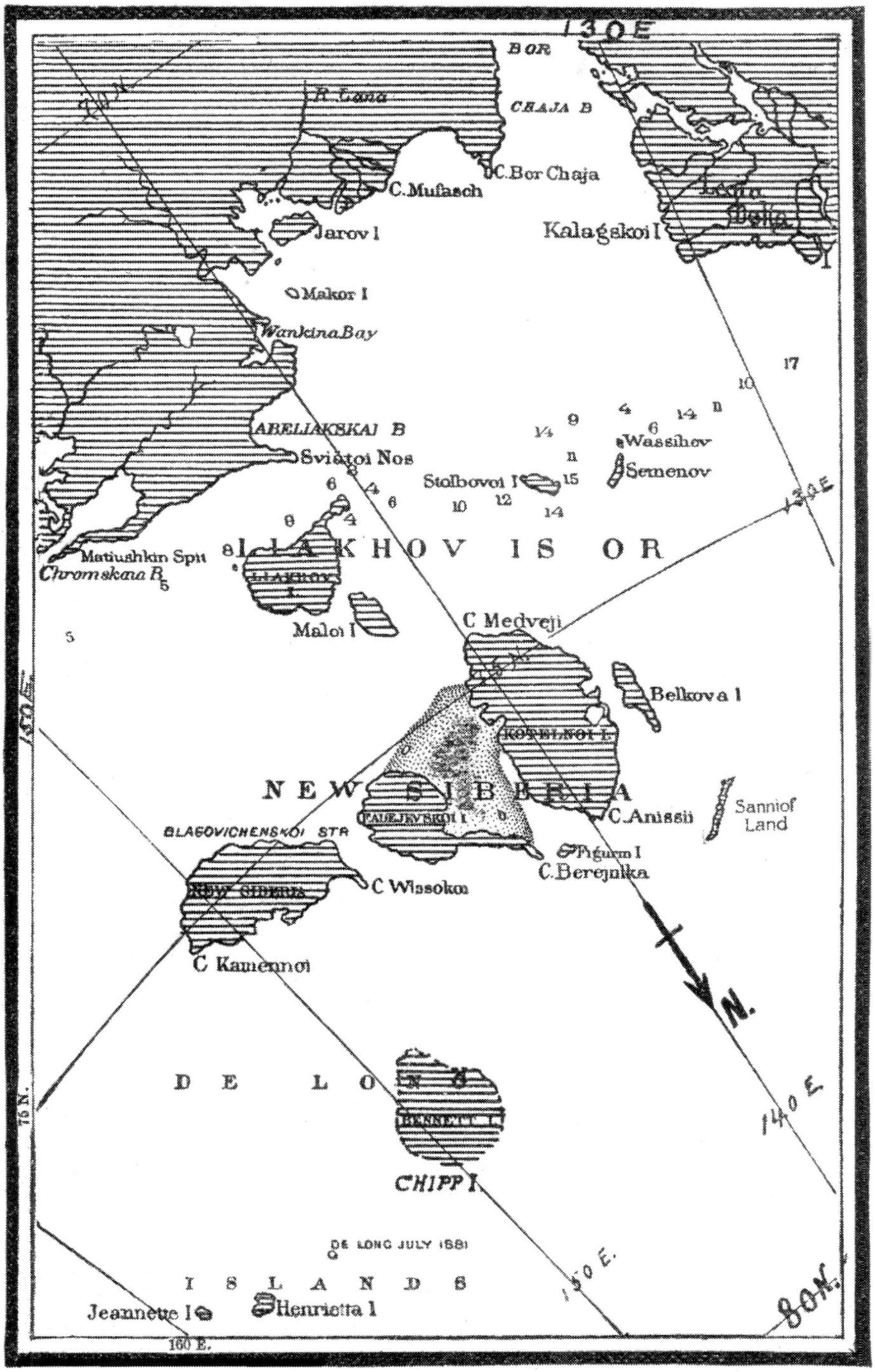

NEW SIBERIAN ISLANDS (Hydrog. Chart, U.S. Navy)

event the large cargoes brought annually for 80 years had not sensibly diminished the supply.

In 1810, Hedenström, leaving the mouth of the Indigirka, reached Nova Sibir in eleven days, and thence attempted to reach very high land (De Long Islands) to the northeast; but after four days' travel met with open water which extended southeast to Bear Islands, some 350 miles. The mainland was reached after 43 days of difficult travelling, the party being saved from starvation only by killing eleven polar bears. Meanwhile Sannikof had returned to Kotelnoi, where he passed the summer with fur and mammoth hunters, and found mammoth and other bones in enormous numbers. Surveying the west coast, he found the cross-marked grave of a Russian hunter, a wooden house partly furnished, and a wrecked vessel. It seems probable that by stress of ice and weather some Archangel hunters had involuntarily made the first voyage around Cape Chelyuskin, and, wrecked on this shore, sealed their discoveries with their lives. Other surveys in 1811, by Sannikof and Pschenizyn, had no important additional results, but from Thaddeus Sannikof saw an island (Bennett) which he could not reach, owing to open water.

It is useless to here detail the remarkable journeys made by Lieutenants F. von Wrangell and P. F. Anjou, in the years 1820–23, on the Siberian Sea. They are full of interest for any reader fond of adventure, but their material outcome was scanty. The New Siberian and Bear Islands were skirted, and every effort under most daring and trying conditions was made to discover land to the east, west, and north. The farthest point reached was in 1823 by Anjou, about 76° 36′ N., 138° W., north of Kotelnoi, where the sea was only 17 fathoms deep: here Anjou saw no sign of the reported and sought-for land of Sannikof, 1811. Extended travel in any direction was impossible on account of open water or unsafe ice, and the excessive roughness of the sea-ice exceeded anything previously experienced, — ice-conditions that are now known to rule on any large water-area subject to high winds or strong currents. WRANGELL thoroughly explored Bear Islands, and sought to discover beyond them the land which ANDREJEV claimed to have seen to the northeast in 1763, and which other explorers had endeavored to find in vain, — since none such exists.

Captain KELLETT discovered, 17th August 1849, the first island to the north of Bering Strait, in 71° 18′ N., 175° 24′ W., named after his ship *Herald*. KELLETT landed with great difficulty, and describes it as granitic rocks, almost perpendicular and nearly inaccessible. In 1855, Commander JOHN RODGERS, U. S. S. *Vincennes*, entered the Arctic Ocean to visit the land 'in about 72° N., 175° W., as placed on the Admiralty charts from the report of H. B. M. S. *Herald*.' RODGERS traversed KELLETT's Plover Land, anchoring, 11th August, in 72° 05′ N., 174° 37′ W., whence a clear view of 30 miles in all quarters showed no land. Later he landed on Herald Island, which has since been visited by HOOPER, *Corwin*, in July, and BERRY, *Rodgers*, in September, 1881. Owing to fog, RODGERS failed to see Wrangell Land, a few miles distant from his track, but he attained 72° 05′ N., 174° 37′ W., the highest latitude then attained by sailing ship, and reaching 176° E., surpassed the farthest of the famous COOK.

The greatest interest and importance attaches to Wrangell Land, which for over a century was believed in, on the strength of Chutchee reports. It is probable that the visits of these natives to the American continent led to a belief in the existence of land north of Cape Jakan, which WRANGELL vainly looked for in clear water from Baranof Rock, 69° 42′ N., 176° 32′ W., 8th April 1824. PETER-

MANN, the great geographer, believed that Wrangell Land extended from the neighborhood of Asia across the Pole to Greenland; and as late as 1881, NORDENSKIÖLD thought it possible that the land might connect with the archipelago to the north of America. The discovery of Wrangell Land is due to an American whaler, THOMAS LONG, who sailed along its south shore in the *Niles*, 14–16th August 1867, when he placed its western extremity in 70° 46′ N., 178° 30′ E., and its southeastern cape in 70° 40′ N., 178° 51′ E. Captains RAYNOR, whaler *Reindeer*, PHILIPS, *Monticello*, and BLIVEN, *Nautilus*, also visited it the same year. Although BLIVEN made its latitude too high, his reports of its east coast tended to confirm the belief that it was an extensive land.

The credit of exploding the theory of a continent belongs to Commander G. W. DE LONG, U. S. Navy, who sailed in 1879 on a voyage of exploration *via* Bering Strait in the *Jeannette*. DE LONG thought Wrangell Land continental, and boldly entered the pack near Herald Island, in 71° 35′ N., 175° W., expecting to reach and winter at Wrangell Land. To DE LONG's dismay the vessel never escaped the pack, and, drifting almost steadily to the westward, passed in sight of and to the north of Wrangell Land, which thus shrank from its assumed continental proportions to that of a small island. Winter came and went; the injured *Jeannette* barely escaped shipwreck from convulsions of the ice-pack, and only avoided foundering through the energy and skill of G. W. MELVILLE, its chief engineer. Another year passed without material change in their condition, save that MELVILLE landed, at risk of his life, 3d June 1881, on a newly discovered (Jeannette) island, in 76° 47′ N., 159° E. It was a barren, rocky, ice-capped islet, with a few dovekies nesting in its cliffs. Another island, Henrietta, in 77° 08′ N., 158° E., relieved the monotony until the crushing of the *Jeannette*, 12th June 1881, in 77° 15′ N., 155° E., left her crew shelterless on the ice-floes, in mid-ocean.

Lieutenant DANENHOWER had long been disabled, Lieutenant CHIPP and three men were sick, and under these conditions DE LONG retreated south to the New Siberian Islands, 150 miles distant. In this fearful journey MELVILLE was full of energy and expedients. Their five boats and nine sledges scarcely carried their 60 days' provisions, which their enfeebled men could not drag in a single load. Added to their misfortunes was a northerly drift that carried them 28 miles to 77° 36′ N., 155° E., the most northerly point ever reached in that sea. An (Bennett) island, 76° 38′ N., 148° E., was soon seen, and landing on it the shipwrecked men rested nine days, exploring 17 miles of the south coast, and obtaining needed supplies of fresh food, the cliffs being alive with birds. Starting 8th August, they landed eleven days later on Thaddeus (Fadejef), one of the New Siberian Islands. In attempting to reach the Lena Delta, CHIPP'S boat, with a crew of eight men, foundered in a storm, 12th September, when DE LONG and MELVILLE were separated. MELVILLE, with nine men, reached, 26th September, a Russian village, Geeomovialocke, through one of the eastern mouths of the Lena.

DE LONG, with Dr AMBLER and twelve men, landed, 17th September, in 77° 15′ N., 155° E., being obliged to abandon his boat owing to shallow water. Armed, loaded with records, food, and other materials, they followed south the barren shores of the Lena, retarded by snow, young ice and disabled men. Delayed till ice could form over unfordable tributaries, game and food failed, 9th October, when one man was dead and others were helpless. Holding fast with AMBLER to the sick and dying men, DE LONG sent two seamen to follow up the Lena in search of relief.

DE LONG and all but three with him died of starvation before 1st November, while the two seamen reached Bulun 29th October, after almost perishing. At the earliest practicable moment, MELVILLE pushed relief measures so energetically that he found, 14th November 1881, the ship's books, when a violent storm prevented farther action. With earliest spring the untiring MELVILLE was in the field, and, 23d March 1882, found the bodies of his unfortunate shipmates.

Apart from valuable physical observations in an unknown region, the geographic results of this expedition were extensive and important. They covered some 50,000 square miles of polar ocean, and clearly indicate the conditions of an equal area between their line of drift and the Asiatic coast. DE LONG proved that the Siberian Ocean is a shallow sea, dotted with islands, and the discovery of the De Long group confirmed the reports of land by SANNIKOF, 1811. The land seen by CHIPP during DE LONG's retreat is probably an outlying island of the group.

Captain C. L. HOOPER, U. S. Revenue Marine *Corwin*, saw Wrangell Land, 10th September 1880; and in 1881, after visiting Herald Island, in 71° 04′ N., 177° 40′ W., landed and explored, with Dr I. C. ROSSE, Wrangell Island, 12th August. A month later, BERRY, in the *Rodgers*, searching for DE LONG, landing in 70° 57′ N., 178° 10′ W., thoroughly explored and mapped the island, which is some 70 miles east and west by 35 broad, with its northernmost point in 71° 32′ N.

The geographic work in connection with the islands of the Siberian Ocean has fittingly closed with a most valuable scientific expedition under the auspices of the Imperial Russian Geographical Society, in charge of Dr A. BUNGE and Baron E. VON TOLL. The year 1885 was passed in Jana Land, where most valuable observations and collections were made. Here, in April, 1886, at Bor-Urjak, 70° 03′ N., on the Dodomo, was found the skeleton of a mammoth covered with moss and other vegetation, the bones mixed with birds' feathers, grass yet green, and other interesting surroundings.

Provided with means for a summer's stay on the islands, TOLL and BUNGE, with eight men, reached Great Liachof Island, where the latter remained for scientific work. TOLL made Kotelnoi his main field-work, but in a journey of 23 days visited Nova Sibir, and later other islands. He traversed the entire coasts of Kotelnoi in 40 days, and was fortunate enough in August to see Sannikof Land (1811), from high land on the northwest of Kotelnoi. TOLL says there were distinctly visible four lofty mountains, extending along the horizon from 14 to 18 degrees west of north, which were united by a foreground of lower land. While he could not see the land (Henrietta Island) of SANNIKOF from Nova Sibir, he thought it might be due to unfavorable atmospheric conditions.

BUNGE meanwhile explored Great Liachof, where the upper layers of clay, above the mixed mud and ice, furnished immense quantities of fossils. In addition to the well-known remains of the mammoth, musk-ox, and rhinoceros, were bones of deer, horses, and two new species of ox. TOLL states that the mammoths were always found in masses of mud and clay pressed into broad fissures of the underlying ground-ice, which forms the base of the low part of the islands. The destruction of the rich fauna at the edge of the ice-masses of the glacial period is attributed to the gradual sinking of the coasts, thus restricting gradually the land area that furnished nutrition.

The contributions made by this expedition are especially valuable to geology, meteorology, botany, and paleontology.

In 1893 the indefatigable Baron TOLL set forth on a paleontological voyage to search to the northeast of the Jana for a well-preserved mammoth. Afterward with Lieutenant SHILEIKO he visited the New Siberian islands, and reaching by dog sledges 75° 37′ N., on the west coast of Kotelnoi, established two provision depots for NANSEN'S possible use. In addition to valuable physical observations and rich collections, TOLL found evidence that in the mammoth period trees grew in 74° N., fully three degrees of latitude beyond their present limit. Geologically this voyage is most important. Returning to the main land, TOLL crossed with reindeer the reported impassable *tundras* from Sviatoi Noss to Dudinka on the Lena, and following up that river reached Yeniseisk 4th December.

SABINE: *Wrangell's Polar Sea*, 2d edition (London 1844); DE LONG: *Voyage of the Jeannette*, 2 v. (Boston 1883); HOOPER & ROSSE: *Cruise of the Corwin, 1881* (Washington 1883); GILDER: *Icepack and Tundra* (New York 1883); BUNGE, TOLL & SCHRENCK: *Expedition nach den Neusibirien Inseln u. Jana-Lande* (St Petersburg 1887).

CHAPTER XIV

SMITH SOUND AND ROBESON CHANNEL

CAPES Alexander and Isabella have been termed the 'Northern Pillars of Hercules,' and not inaptly so, because these ice-bound rock-masses mark the limits of safe and dangerous ice-navigation, and since also between these portals have passed the expeditions that have again and again pushed poleward the Ultima Thule of the world. In his unequalled Arctic voyage of 1616 (page 20) BAFFIN attained a point within 25 miles of Cape Alexander, and discovered Smith Sound, running to the north of 78°. It was 236 years before BAFFIN'S latitude, in a tiny shallop, was surpassed; and then, although with steam, by only a few miles. The northern limits of Smith Sound were determined by Captain E. A. INGLEFIELD in the *Isabel*, in a summer search for FRANKLIN, and to ascertain if Smith Sound was connected with the Polar Sea.

At midnight, 26th August 1852, the *Isabel* was off Cape Alexander, the farthest seen by BAFFIN, and the northward view opened to the long expectant INGLEFIELD, who 'beheld the open sea stretching through seven points of the compass . . . . bounded on the east and west by distant headlands (Cairn Point and Cape Albert).' Twelve hours later the *Isabel* was turned back by ice from 78° 21′ N., — hardly 40 miles beyond BAFFIN, — and reached England in November. Considering expenditure of time, money and effort, this is one of the most success-

No VIII.

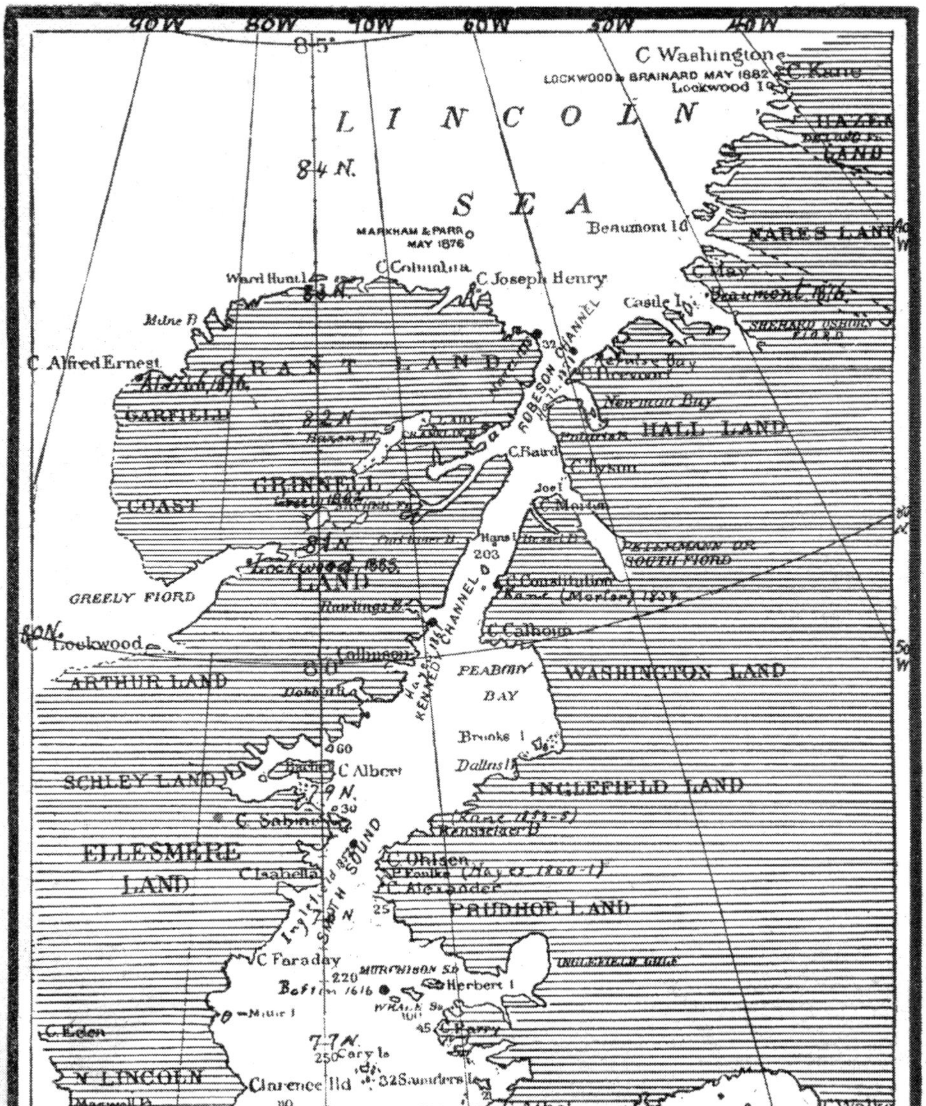

SMITH SOUND AND N. W. GREENLAND

(Hydrog. Chart, U.S. Navy)

ful of modern voyages. INGLEFIELD laid down 600 miles of uncharted coast, rectified many errors, outlined Smith, and penetrated far into Jones' Sound.

American energy, liberality and courage were destined to win the highest laurels by the new polar passage, now known as the American route. ELISHA KENT KANE first passed north of Smith Sound into the enclosed sea that appropriately bears his name. With the *Advance,* fitted out in 1853 by HENRY GRINNELL and GEORGE PEABODY, KANE contemplated a search for FRANKLIN and a northerly extension of INGLEFIELD'S discoveries. In August the *Advance* had beaten her steam-rival, *Isabel,* and was off Littleton Island, where, detained by ice, KANE prudently and most fortunately cached a boat and provisions, at Life-Boat Cove. Alternately favored and harassed by ice-conditions, KANE rounded Cairn Point and was fairly in Kane Sea when a gale almost wrecked the *Advance.* Disregarding the cautious and almost unanimous advice of his officers to seek a harbor to the south, KANE warped the brig to the east, where she found her final moorings, in Rensselaer Harbor, 78° 37′ N., 71° W.

An autumnal sledge journey opened up the high coast at Cape Constitution, while other parties examined the great ice-cap of Greenland, and cached sledging supplies under the towering face of Humboldt glacier. Winter killed their dogs, scurvy enfeebled the ill-fed crew, and spring opened with excessive cold; but KANE firmly adhered to his plans. An unfortunate sledge journey, in March 1854, caused the death of two men, disabled two others and the rest only escaped by KANE'S heroic personal efforts. In May, his surgeon, ISAAC I. HAYES, crossed Kane Sea, with one man and a dog-sledge, and, first of all explorers, reaching Grinnell Land, traced it to Cape Frazer, 79° 43′ N. On the Greenland side, mate

William Morton and Hans Hendrik skirting Humboldt glacier scaled, 24th June 1854, the south side of Cape Constitution, 80° 35′ N., whence from a height of 500 feet they saw the southwest half of Kennedy Channel, perfectly ice-free, and the east coast of Grinnell Land, probably to Mount Ross, 80° 58′ N., 75 miles beyond Hayes's farthest.

Realizing his desperate situation Kane vainly attempted, in July 1854, to reach Beechey Island, 400 miles distant, to obtain assistance from Belcher, who was then preparing to abandon his own ice-bound ships. Fortunately the ice-conditions prevented Kane from even reaching Cape Parry. On 28th August, Hayes, believing with the majority of the crew that another winter would be fatal, left the *Advance* with Kane's consent, in a futile and desperate autumnal attempt to reach Upernivik. After terrible sufferings, wherefrom they would have perished except for the Etah Eskimo, the party returned to the ship in December.

The spring of 1855 found the party in a deplorable condition from unfit diet, increasing scurvy and mental depression. Several men were completely disabled, the harbor ice was solid, fuel exhausted, the upper works of the brig were burned, and their friends the natives had been reduced by failure of game to the lowest stages of emaciation. There was no other course but to abandon the brig (which was done, 20th May 1855), and attempt to reach Upernivik by boat.

By indefatigable exertions of Kane, supplemented by Eskimo contributions of transportation and food, the invalids, records, and boats were transported 80 miles to open water at Cape Alexander; but the journey was unfortunately marked by the death, caused by over-exertions for the common safety, of their ship's carpenter, the

Dane OHLSEN, in sight of the open water that was to bring safety to his comrades. Fifty days brought KANE'S party to Upernivik, and they returned to the United States in the squadron under Lieutenant HARSTENE, who had been sent out for their rescue.

The results of KANE'S expedition were most important. It extended northward Grinnell Land from Bache Island to Carl Ritter Bay, Greenland from Cape Ingersoll to Cape Constitution, and outlined free-water ways that have been more persistently and safely followed poleward than any other. The scientific contributions were most interesting, especially those relating to the flora, fauna and ethnography of extreme western Greenland, — a region isolated by the great surrounding ice-cap. The tidal, meteorological, magnetic and glacier observations were most valuable, not only by their remoteness from others, but also as forming the basis and stimulus of the existing magnificent series of physical contributions relating to West Greenland.

Immediately after the return of KANE, his surgeon ISAAC ISRAEL HAYES succeeded, through the advocacy of Professors BACHE and HENRY and the support of GRINNELL, in fitting out a new expedition. Failing to obtain the steam vessel that he recognized as necessary for sure success, HAYES sailed on the schooner *United States*, and on 4th September 1860 entered Port Foulke, 78° 18′ N., 73° W. Here he wisely decided to winter, recognizing that any sailing vessel pushed into Kane Basin is hopelessly situated. That autumn he traversed the inland ice 40 miles, to that time the most successful attempt to penetrate the interior of Greenland.

The winter proved unfortunate; his Eskimo dog-driver, PETER, fled to the inland ice and perished; his sledging-dogs, the mainstay of future explorations, died. A greater calamity was the death of SONTAG, the astronomer, whose scientific acquirements and previous Arctic experience with KANE made him the most valuable member of the party. SONTAG's death by freezing occurred on a sledging trip with HANS HENDRIK to communicate with the Etah Eskimo, with whom HANS had lived from the end of the Kane expedition in 1855. SONTAG's fate gave rise to insinuations as to HANS's connection therewith, but there are no grounds for imputing anything more than lack of judgment and decision under calamitous circumstances that would have sorely tried any man.[1]

In the spring of 1860 HAYES, securing the co-operation of the natives, visited Van Rensselaer Harbor, where he found the dismantled *Advance* of KANE vanished and its mooring place filled with ice mast-high. Later HAYES commenced his northern journey, and, imbued with his theory of an open polar sea, transported a boat to secure his retreat. After losing four weeks' time and utterly exhausting his men, he sent back the boat, proceeded by dog-sledge, and with three men reached Cape Hawks 11th May, having been 39 days on a journey that he could have made in ten days with dogs alone.

Four days later JANSEN, being disabled, was left with McDONALD, while HAYES and KNORR pushed on, reaching their farthest, 19th May 1861, at a point called Cape Lieber, which HAYES placed by his unreliable observations in 81° 35′ N., 70° 30′ W. Stopped by water-holes, an unobstructed view was had from a height of 800 feet. HAYES says, 'The sea was a mottled sheet of white and black patches multiplied in size as they receded until the belt of the water-sky blended them together. . . . All the evidences showed that I stood upon the shores of the

[1] See *Memoirs of* HANS HENDRIK, *written by himself*, translated by Dr. H. RINK (London 1878).

polar basin, and that the broad ocean lay at my feet; in dim outline [was] a noble headland, judged to be in 82° 30′ N. . . . There was no land visible except the coast on which I stood.'

Subsequent expeditions have surveyed this region with great accuracy, and it is known that HAYES neither saw an 'open polar sea,' nor reached the astronomical point mentioned; for there is an error either of one and a half degrees of latitude or six and a half degrees of longitude. HAYES left uncharted the North Greenland coast, and elsewhere says: 'No land to the eastward. As it would not have been difficult through such an atmosphere to see a distance of 50 or 60 miles, it would appear that Kennedy Channel is wider than heretofore supposed.' Kennedy Channel is only 30 miles wide, and the lofty Greenland coast is plainly visible from every point of the Grinnell Land shore in equal latitude. From the north side of Rawlings Bay, Greenland to the north of Cape Constitution plainly comes into view. As HAYES's map of Greenland ends with that cape, he could not have reached the north side of Rawlings Bay, and his farthest could not have been beyond Cape Joseph Goode, 80° 11′ N., which point fulfils best the varied conditions imposed by the scientific journals, personal narrative and maps of HAYES, and the topography of the region. The 'open polar sea,' an endless source of theory and discussion, was only the south half of Kennedy Channel, which freezes late and opens early owing to its very high tides, some 30 feet; it was found open by MORTON, and has never been seen entirely closed.

Despite the mythical 'open polar sea' and his failure to exceed the latitude of MORTON, HAYES's adventurous journey contributed to Arctic geography. He was the first civilized man to visit the lands of Ellesmere and

Grinnell, whose coasts he surveyed between 77° and 78° N., adding Hayes Sound, Bache Island and other details to our maps. In 1869, HAYES visited Greenland with the artist BRADFORD in the *Panther*, his voyage contributing to our knowledge of the glaciers of Melville Bay; otherwise it was devoid of interest.

Another American here ventured his fortunes, having gained earlier reputation and experience in the Franklin search. Though not professionally qualified to command his ship, nor to assume charge of the scientific work connected with his expedition, yet CHARLES FRANCIS HALL had the qualities that mark the genuine Arctic explorer. Fertile in resources, indefatigable in exertions, sparing no personal effort or exposure, faithful in record and conservative in action, great results were expected from HALL in his voyage in the *Polaris*, a reconstructed tug of the United States Navy unsuited for Arctic work.

HALL was uncertain whether he would follow Smith or Jones Sound, but Smith Sound being entirely ice-free the question of route was settled. With phenomenal good fortune the *Polaris* crossed the parallel of Van Rensselaer Harbor, and steamed uninterruptedly through Kane Basin, Kennedy Channel, Hall Basin, and Robeson Channel into the hitherto inaccessible Polar Ocean. She was stopped, by a heavy pack, 30th August 1870, in 82° 11′ N., the highest northing then attained by vessel, within 34 miles of PARRY's latitude by boat north of Spitzbergen; in this sea it was 200 miles beyond KANE's *Advance*.

From his highest point HALL saw that the adjacent coast of Greenland trended east, while the opposite shore of Grinnell Land continued to the north.

Contrary to the rule of ice-navigation HALL sought, with unfortunate results, a (Repulse) harbor on an east shore. The *Polaris*, unable to enter, was caught by the main polar pack, and, after 50 miles' southing, was forced on the Greenland shore. Barely escaping destruction, she obtained anchorage under an enormous floeberg, 650 by 450 by 300 feet in size, in 81° 37′ N., 62° W. In this open roadstead, called Thank God Harbor, the expedition wintered.

While preparatory work for winter was in progress, HALL made a sledge journey northward to Cape Brevoort, 82° N., whence he saw 'land extending on the west side of the [Robeson] strait to the north, as far as we can discover, about 83° 05′ N.;' subsequent observations show that HALL was very nearly correct. Indirectly the journey proved fatal to HALL and the expedition, for immediately on his return he was taken violently sick and died, 8th November 1871.

Desultory and unsuccessful efforts to go north by boat were made in 1872; a dog-sledge trip was made to the south by Dr BESSELS, to Petermann Fiord, and Sergeant F. MEYER, signal corps, U. S. Army, reached on foot Repulse Harbor, 82° 09′ N., to that time the most northerly land ever attained by civilized man.

In returning south that autumn, another ice-rule was ignored by the *Polaris* with disastrous results. Pushed into an impassable ice-pack, she was anchored to a floe, and left to the mercy of ice, wind and current. For two months the ship drifted slowly southward, and was off Northumberland Island, when a violent gale disrupted the pack and nearly destroyed her. Part of the terror-stricken crew, escaping in the darkness to the ice-pack, experienced the horrors of a mid-winter ice-drift, whose appalling dangers and bitter privations can scarcely be appreciated. Five months later, after a drift of 1,300 miles, the despairing party were picked up by the *Tigress*, off Labrador, 30th April 1873, not only unreduced in numbers, but with a girl baby born to the Eskimo, HANNAH.

Those who held to the *Polaris* fared better. The ship drifted to land, near Life-Boat Cove, of KANE, where she was beached, being unseaworthy, and a shelter (Polaris House) constructed. In 1873, boats were built; and, starting on a journey to Upernivik, the party was rescued at Cape York by the whaler *Ravenscraig*, June 22d.

The geographic results of this expedition were extensive and valuable. HALL had completed the exploration of Kennedy Channel, discovered Hall Basin and Robeson Channel, and even visited the hitherto inaccessible polar ocean. He had extended both Grinnell Land and Greenland northward nearly two degrees of latitude, the former practically to its extreme limits, and explored extensive portions of the latter. The most important physical discovery of HALL was the termination of the northwestern inland-ice of Greenland at Petermann Fiord, and the entirely ice-free condition of several thousand square miles of the most northerly part of Greenland, — the largest known area of bare ground of that continent.

The unprecedented successes of HALL stimulated anew the interest of Great Britain in Arctic discovery, and a large well-found expedition, the *Alert*, *Discovery* and *Valorous*, sailed in 1875. The command was entrusted to Captain GEORGE NARES, one of the ablest of British navigators, who was detached from the celebrated *Challenger* expedition for this important service. The *Valorous* returned from Disco, while the other ships reached Cape Sabine without difficulty, wisely caching 3,600 rations on the southeast Cary Island to ensure possible safe retreat; 3,600 rations were later cached at Cape Hawks, 1,000 at Cape Lincoln, and smaller amounts at other important points.

The journey northward from Cape Sabine was as unlike HALL's as possible, — a constant struggle against adverse ice-conditions instead of an open sea, NARES following the sound principles of hugging the west shore, keeping free from the heavy pack, and following shore channels opened by tide or wind. His squadron passed Cape Lieber, and crossing Lady Franklin Bay, 25th August 1875, found a fine, land-locked (Discovery) harbor, where he put his supporting ship, *Discovery*, in winter quarters. Pushing his own ship, *Alert*, northward along the west shore of Kennedy Channel, he was able to reach Floeberg Beach, 82° 25′ N., 62° W., the most northerly point ever reached by ship; there he wintered on the exposed shores of the Polar Ocean. Autumnal sledging parties laid out depots for the next spring, at fearful cost, as eight men were frost bitten, three so badly as to render amputations necessary. Lieutenant PELHAM ALDRICH was more fortunate in an exploring trip wherein he exceeded the hitherto unsurpassed latitude of PARRY, 1827, by reaching 82° 48′ N., on the coast of Grinnell Land, which stretched yet farther to the north beyond the 83d parallel. The sun was absent 145 days, and the average cold of the winter was intenser than any before experienced, but spring found the crew of the *Alert* in health. Communication was opened with the *Discovery*, costing the life by frost-bite of CHRISTIAN PETERSEN, despite the heroic exertions of Lieutenants RAWSON and EGERTON.

With returning sunlight NARES commenced the sledging work for which he was sent forth. His efforts were divided between a direct journey toward the North Pole over the frozen sea, which was made by MARKHAM (page 171), and the exploration of the north shores of Grinnell Land by Lieutenant ALDRICH, the latter very successful, as he traced 220 miles of new coast. Passing Cape Columbia 83° 07′ N., ALDRICH reached Cape Alfred Ernest, 18th May 1876, in 82° 16′ N., 86° W., where Grinnell Land trends to the southwest. Attacked by the scurvy on the return trip, only ALDRICH and one man out of eight were able to haul, and all would have perished save for Lieutenant MAY, who came to aid them.

From the *Discovery*, Lieutenant ARCHER surveyed thoroughly a (Archer) fiord to the south, while Dr COPPINGER visited Petermann Fiord, after supporting Lieutenant L. A. BEAUMONT, who explored the Greenland coasts. BEAUMONT travelled *via* the *Alert* to Repulse Harbor, and followed the coast to Cape Bryant. Pushing across Sherard Osborn Fiord, he found his men to be weakened by scurvy, while deep, soft snow made dragging very heavy. Leaving his party to recuperate in camp, BEAUMONT with one man succeeded, 20th May 1876, in reaching the eastern shore, 82° 20′ N., 51° W. Disease steadily disabled his sledge-men, and although he dropped instruments, food and everything not indispensable, yet hope almost failed when he found the ice of Robeson Channel too rotten to permit crossing to the *Alert*, at Floeberg Beach. All but one of his eight men were unable to work, and only the advent of Lieutenant RAWSON and Dr COPPINGER saved the gallant party. As it was, two men, PAUL and HAND, died at Thank God Harbor, HALL's old quarters.

As his sledge parties had all broken down, and 36 cases of scurvy had occurred on the *Alert* alone, NARES wisely decided to return. By daring and skilful ice-navigation he successfully extricated his ships and reached England that autumn (1876).

The geographic work done was extensive and important. It did not consist alone in carrying a British ship and in planting the Union Jack in a higher latitude by land and by sea than had ever before been attained. The coast of Grinnell Land was surveyed from the head of Archer Fiord northward to Cape Columbia and thence to its probable western limit. The north shores of Greenland had been extended from Cape Bryant to Cape Britannia, and BEAUMONT'S observations had determined that the land to the eastward of Sherard Osborn Fiord was equally ice-free with that seen by HALL. The tidal, magnetic and meteorological observations were valuable additions to the physical sciences, and the tenacious courage and persistent energy of its men and officers added new laurels to the British navy.

Americans again sought the dangerous waters of the West Greenland Channel, engaged neither in an attempt to reach the 'Open Polar Sea,' nor to attain the north geographic pole. The Lady Franklin Bay expedition under Lieutenant A. W. GREELY occupied one of the International Circumpolar stations, for systematic scientific work, and as such its fortunes and successes are considered in a subsequent chapter.

BAFFIN'S *Voyages* (see page 21); INGLEFIELD: *Summer Search for Franklin* (London 1853); KANE: *Second Grinnell Expedition*, 2 v. (Philadelphia 1856); HAYES: *Arctic Boat Journey* (New York 1860), and *Open Polar Sea* (New York 1867); DAVIS: *Polaris [Hall's] North Polar Expedition* (Washington 1876); BLAKE: *Arctic Experiences; Tyson's Drift* (New York 1874); NARES: *Voyage to the Polar Sea*, 2 v. (London 1877); BESSELS: *Die Amerikanische Nordpol Expedition* (Leipzig 1879).

CHAPTER XV

FRANZ JOSEF LAND

PETERMANN says: 'I consider it highly probable that the great Arctic pioneer, WILLIAM BAFFIN, may have seen the western shores of Franz Josef Land in 1614.' However that may be, our present knowledge of this Arctic archipelago is due to the exertions of the Austrian soldier PAYER, and the English yachtsman LEIGH SMITH.

At the instance of Lieutenant CARL WEYPRECHT, Austrian-Hungarian navy, Count WILCZEK fitted out two expeditions to explore the Nova Zembla Sea and try the northwest passage. With WEYPRECHT was associated Lieutenant JULIUS PAYER, Austrian-Hungarian army; and in a preliminary journey in the *Isbjörn*, they reached, 1st September 1871, 78° 48′ N., 42° E. The *Tegetthof* carried the second expedition, WEYPRECHT exercising marine command and scientific control, while all land explorations fell to PAYER.

The *Tegetthof*, with 24 souls and eight dogs, fell in with WILCZEK near Barents Isle, where with the *Isbjörn* he was placing supplies for their possible retreat. Ice conditions parted them, 20th August 1872, when the *Tegetthof*, steaming north, was beset the same day in sight of Nova Zembla, in 76° 22′ N., 63° E. Destined never to reach open water again, the ship drifted with the ice-pack here and there as the winds blew; for in that sea the wind con-

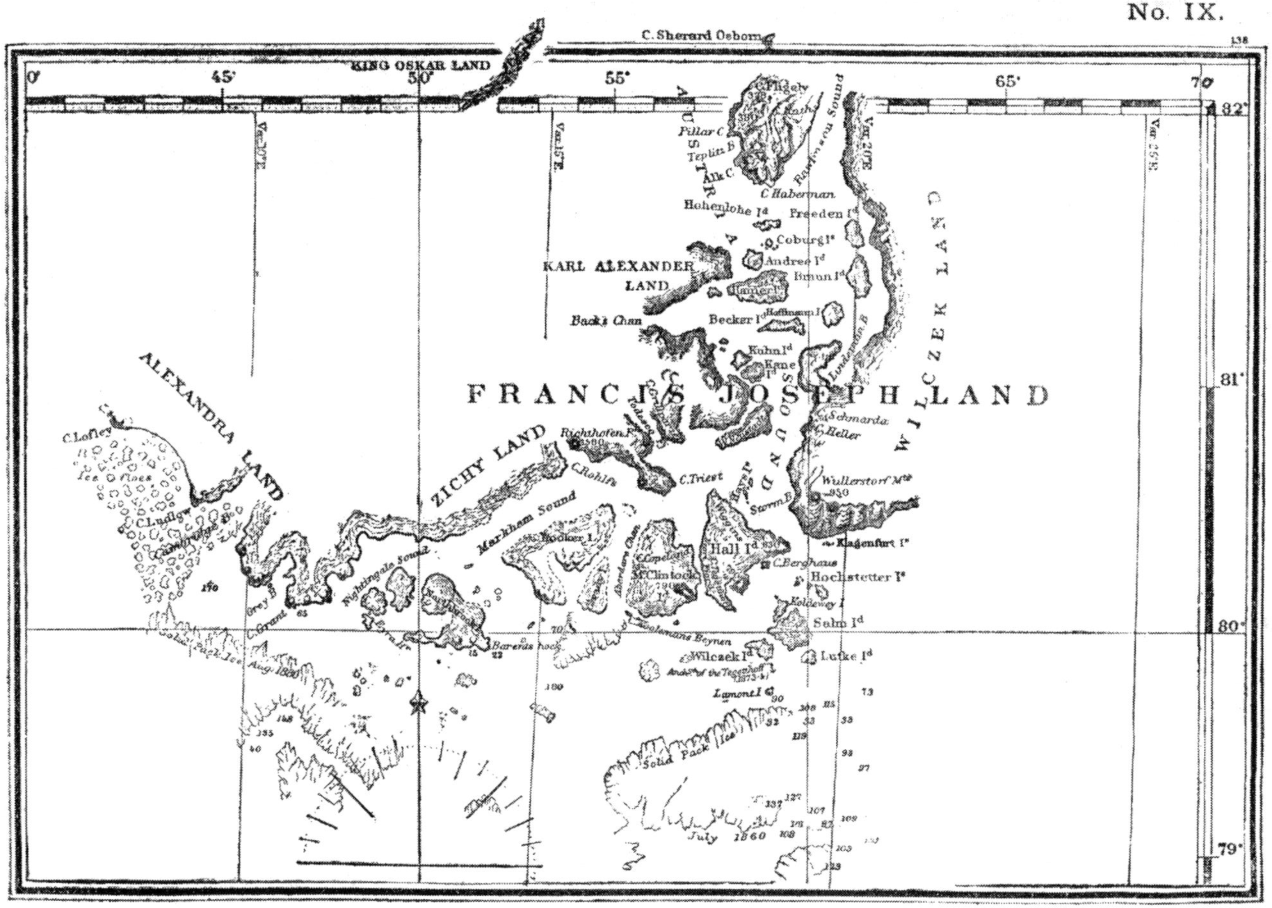

FRANZ JOSEF ARCHIPELAGO (HYDROGRAPHIC CHART, U.S.N.)

trols ice-movements. From week to week the bleak but welcome shores of Nova Zembla faded gradually, until at last only wastes of ice formed their narrow horizon.

The terrible conditions under which they existed may be surmised from PAYER'S narrative: 'On 13th October 1882, our floe broke across immediately under our ship. . . . We were surrounded and nipped. . . . Our floe was now crushed, and its blocks, driven hither and thither, towered fathoms high above the ship, and forced the massive oak timbers against the hull. The ice under the ship began to raise her, and we made ready to abandon her, if, as seemed inevitable, she should be crushed. . . . The *Tegetthof* heeled over, and huge piles of ice threatened to precipitate themselves on her, but the pressure abated.'

Driven by fearful ice-convulsions to contemplate the loss of their ship at the beginning of an Arctic winter, they built a house on the main floe, where coal, fuel, and other supplies were stored. This done, they applied themselves to observations, exercise, short journeys, and an occasional bear-hunt.

Thus passed the first winter; and when the returning sun cast its first rays on the haggard crew, they felt that the least educated man of all expressed the inmost feelings of even the most cultured when he said, 'Blessed sunlight!' With the sun came bears, in such numbers as to contribute materially to their food supply, and with such ferocity as to make their visits dangerous. From February the ship drifted first northwest and then north, attaining its greatest longitude, 71° E., in 79° N.; with summer they moved slowly westward, reaching, 8th July, 59° 05′ E., nearly their most westerly point. It developed that their drift was the resultant of southwest winds, as affected by the presence of (Franz Josef) land to the north.

Summer was marked by no material change in the ice, but the presence of birds, seal, and bears insured an ample supply of fresh meat, thus preserving health. The monotony was broken 30th August 1873, when the rising mist revealed at mid-day, far to the northwest, the outlines of a bold and rocky land. It was hailed with transports of enthusiasm, for they realized that their toil and suffering had not been all in vain, since they had added a new land to the known domain of the world. It was the end of September before they dared to leave the ship, then in 79° 58′ N., and by a forced march unsuccessfully endeavored to reach an (Hochstetter) island. A fog fell as they marched, and only by the sagacity of their dogs did they return safely to their ship. With stable ice, incident to coming winter, they were more fortunate in November, and made short trips to Wilczek Island.

The second winter passed quietly, although they were harassed by constant fears that the drift would carry them away from the unexplored land. Their feverish impatience is best illustrated by PAYER, who says: 'The reappearance of the sun last year was tantamount to a deliverance from hell itself; but now the sun was nothing to us save as a means to an end. Would it enable us to begin our sledge-journeys?' In March 1874, PAYER, with six men and three dogs, visited Hall Island, a plateau land 2,000 feet high. A bear was killed during the five days' absence; but the intense cold, 59° below zero, frosted the men so badly that they had to return. The main journey began 26th March, with ten men and three dogs. Violent blizzards, intense cold and rough ice discouraged them, but consolation came in bagging four polar bears. Entering a sound (Austria) 1st April, they camped a week later at Hohenloe Island, 81° 37′ N., with two men worn out. Leaving them under a third man, PAYER started north with the rest, full of hopes which the first day threatened to destroy entirely.

While crossing Middendorf glacier, a snow-bridge gave way, and sledge, dogs, and man fell 30 feet down the crevasse. PAYER was dragged in harness to the very edge, he being saved by the sledge wedging, while the man, ZANINOVICH, was thrown on an ice-ledge, and the dogs held fast in their traces. The only chance of rescue lay six miles distant, in the party left at Hohenloe Island. Throwing off all outer garments, PAYER ran in stocking feet, through deep snow, this distance in an hour. Such despatch was made that the relief party reached the crevasse four and a half hours after the accident, and lowering a man, drew up ZANINOVICH, dogs, and sledge uninjured.

PAYER reached, 12th April 1874, Cape Fligely, 82° 05′ N., 58° E., which was, and yet remains, the highest land ever attained in the Old World. Numerous water-holes and rotten ice forbade farther advance, as there was a very large area of clear water to the north and west. From a height of 1,000 feet it was seen that the land was part of an archipelago, as extensive as Spitzbergen. PAYER says: 'Rudolf Land still stretched [from Fligely] in a north-easterly direction to Cape Sherard Osborn. Blue mountain-ranges lay in the distant north, indicating masses of land. These we called King Oscar Land and Petermann Land; the mountainous extremity on the west of the latter lay beyond the 83d degree of north latitude.'

The return to the ship was peculiarly difficult owing to increasing water-holes, but their fears of the pack breaking up during their absence were only too groundless. The ice remaining fast, the *Tegetthof* was abandoned 20th May. The ruggedness of the floes was such that after 'two months of indescribable efforts,' says PAYER, 'the distance between us and the ship was not more than two

German miles.' Persevering, they reached the free sea, 15th August 1874, and nine days later fell in with Russian fishermen on the Nova Zembla coast.

PAYER was followed by LEIGH SMITH, steam-yacht *Eira*, who in 1880 unsuccessfully endeavored to reach Jan Mayen, East Greenland, and North Spitzbergen. SMITH then turned toward the coast of Franz Josef Land, which had remained unvisited since 1874, although DE BRUYNE, in the *Willem Barents*, sighted its high land, probably Northbrook Island, 7th September 1879. Picking up the pack in 77° 10′ N., 40° E., the *Eira* was driven by ice and weather this way and that, and on 14th August was discovered a small (May) isle southwest of McClintock Island. SMITH pushed his discoveries with such judgment and energy as to cover the whole coast of Franz Josef Land from 42° E. to 54° E., the most westerly point of the south shore seen by PAYER. From McClintock Island to Cape Neale, the southwest point of Cambridge Bay, the whole fringe of outlying islands, as well as the main shore, were surveyed. This included Brady, Hooker, Northbrook, and many smaller islands, Nightingale and other sounds, and a number of bays. A secure (Eira) harbor was found in 80° 04′ N., 48° 40′ E. From the most northerly point, 80° 19′ N., 44° 52′ E., it was clearly seen that the western coast, to which the name of Alexandra Land was given, trends decidedly to the north-northwest, from Cape Ludlow to Cape Lofley, in about 81° N. Franz Josef Land was followed to the east as far as Wilczek Island, where, 30th August 1880, open water marked the former location of the beset *Tegetthoff*, 1874; and thence the *Eira* proceeded to Spitzbergen. Every opportunity was improved to collect specimens and make observations; valuable botanical and geological collections were made on land, and of invertebrates in the sea.

Trying his fortunes the following year, SMITH arrived, 28th June 1881, in sight of Cape Ludlow; but unfavorable ice-conditions drove him to the east to Cape Flora, Northbrook Island, where he lost his ship. The *Eira* sank within two hours after the leak was discovered, but by great energy the boats and a considerable quantity of supplies were saved. Supplementing his provisions by hunting, SMITH carried his party through the winter quite comfortably. Taking to their boats the middle of June, by six weeks' strenuous efforts they reached the open sea. Following southward the west coast of Nova Zembla, the shipwrecked mariners fell in with the *Willem Barents*, 3d August 1882, near Matthew Strait, in advance of the arrival of the *Hope*, sent out by the British government.

SMITH's voyages were most important, for they not only extended far to the northwest the limits of Franz Josef Land, but his observations and experiences disclosed the comparative richness of its fauna and flora. They brought forward this region prominently as a suitable base whence extended journeys could be made safely to the northward by any suitably equipped and well-led party that might be landed on its shores, and from which a boat retreat to Nova Zembla is also practicable.

The very important hydrographic work and physical observations of the Norwegian North-Atlantic expedition, 1876–78 (MOHN *et al.: Den Norske Nordhavns-Expedition 1876–1878*, Christiana, 1883) in the Spitzbergen and Barents sea stimulated farther scientific research in Arctic waters. Holland sent out for many summers, from 1878 on, the *Willem Barents* to work especially in Barents sea. The success of the various able officers who have commanded have been most gratifying and important. They extended their investigations almost to the very shores of Franz Josef Land, and but for fog doubtless other points of the archipelago would have been seen besides that by DE BRUYNE in 1879 (p. 199).

Little doubt exists that the Franz Josef archipelago extends westerly through a line of islands, to which White Island and Wiche Land belong, to the Spitzbergen group. In 1884 Captain KJELDSEN discovered a small island, which, seen again by Captain SORRENSEN, 28th August 1884, was named White Island. In 1887 Captain E. H. JOHANESEN discovered to the east of Northeast Land an island, some 2,000 feet high, in 80° 10′ N., 30° 03′ E.

Mr E. F. Jackson landed, 7th September 1894, near Cape Flora, Franz Josef Land, erected two houses and wintered. His ship, *Windward,* by a voyage of sixty-five days, through an ice-pack three hundred miles wide, reached Vardö 10th September, 1895, twelve of her crew sick with scurvy and two dead. Having established advance depots as far as 81° 20′ N., Jackson started north in May, 1895, and it is hoped will outline the northern limits of this archipelago.

WEYPRECHT: *Sulla spedizione polare austro-ungarica* (Trieste 1875); PAYER: *New Lands within the Arctic Circle*, 2 v. (London 1876). See *Petermann's Geographische Mittheilungen,* xx and xxi (1874–75). For SMITH'S Voyages, see *Proceedings Royal Geographical Society* (1881 and 1883). DE BRUYNE *et al.; Verslagen omtrent den* (1st–8th) *tocht v. d. Willem Barents, 1877–1885* (Haarlem 1879–1886).

CHAPTER XVI

THE INTERNATIONAL CIRCUMPOLAR STATIONS

THE importance of scientific research in the Arctic regions has been more or less appreciated since the early days of the 19th century. Only within the last 30 years, however, have the natural sciences been fully represented on polar voyages, and valuable as were the former individual contributions, yet they were restricted and inconclusive. A revolution was wrought in this direction through the efforts of Lieutenant CHARLES WEYPRECHT, Austrian navy, which eventuated in the establishment of the International Circumpolar stations.

His experiences in the *Tegetthof* (see Chapter XV) bore fruit in an address before the German Scientific and Medical Association of Gratz, in 1875. Demonstrating that extensive Arctic explorations were essential to the full elucidation of the laws of Nature, he urged that scientific methods should dominate future plans and action. Scientific investigations should invariably be the primary object, and geographic discoveries — since thus they were alone of decided value — should be attempted in directions where they would extend the fields of scientific inquiry. The subjects to be investigated should decide the locality of observing stations, and the series of observations should be simultaneous, co-operative and continuous.

A commission of eminent scientists, appointed by

Prince BISMARCK, reported that the work was of great value, that the united action of several nations was essential to complete success, and recommended it strongly to the Bundesrath and to all nations interested in science.

WEYPRECHT and Count WILCZEK drew up a plan of the work, which was submitted to the International Meteorological Congress, and received its decision 'that these observations will be of the highest importance in developing meteorology and in extending our knowledge of terrestrial magnetism.' From this recommendation arose the International Polar Conference, at Hamburg, 1st October 1879, of which Dr NEUMAYER was president, eleven nations being represented by delegates or by favorable communications pledging support. The second conference met at Berne, 7th August 1880, when the schedule of optional and obligatory observations, previously drawn up, was adhered to.

Eventually 15 expeditions were sent forth: Denmark, Germany, Russia and the United States each occupied two stations; Austria-Hungary, Finland, France, Great Britain, Holland, Norway and Sweden established one each. Thirty-four permanent observatories (among them Peking, Shanghai, Rio de Janeiro, Munich, Tiflis, and Bombay), adopted the scheme of observations, thus raising the number of co-operating stations to 49, and making it a most notable instance of international scientific action.

The Austrian-Hungarian expedition was sent at the expense of Count WILCZEK, under the command of Lieutenant EMIL VON WOHLGEMUTH, A.-H. Navy, in the *Pola*, which failing to make a landing in May, returned to Tromsöe. Renewing the attempt in June, Jan Mayen was sighted by the *Pola* on the 27th; but it was not until 13th July 1882, that she could make a harbor, and later

No. X.

ARCTIC REGIONS, SHOWING LOCATION OF
CIRCUMPOLAR STATIONS, 1881–83

she was twice driven back to it for shelter. The station was finally established at Mary Mussy Bay, 70° 00′ N., 8° 28′ W., and occupied till the return of the *Pola*, 4th August 1883. WOHLGEMUTH not only carried out his complete series of observations, but by boat and land journeys explored and charted the whole island, geographical work badly needed.

The regular Danish station was established under Professor A. F. W. PAULSEN, at Godthaab, Greenland, 64° 11′ N., 51° 40′ W., and occupied from 1st August 1882 to 31st August 1883. Special observations were made in atmospheric electricity, the march of temperatures in rock, earth and air, and measurement of the parallax of auroras, wherefrom the comparatively small altitude of the phenomena was determined.

The primary object of Lieutenant A. P. HOVGAARD, Danish steamer *Dijmphna*, was the discovery of new lands, but his secondary aim — co-operation with the international stations — dominated from his besetment in the Sea of Kara, 71° N., 64° E., 18th September 1882. She wintered in company with the *Varna*, supplementing the ordinary observations by others on the action of chlorides and salts in the ice, salinity and temperature of different layers of the sea and the action of the pack, which moved with the prevailing wind. The *Dijmphna* drifted here and there, and after losing her screw fortunately escaped without shipwreck, westward through Waigat Strait, 25th September 1883.

Finland opened a station at Sodankyla, 67° 24′ N., 26° 36′ W., through the exertions of Professor SELIM LEMSTRÖM, of Arctic experience, who initiated the work and then turned it over to ERNST BIESE. The station was occupied from 29th August 1882 to 1st September 1883; in addition to its observations, LEMSTRÖM produced auroral displays electrically. In connection with this station and the one at Bossekop there was maintained in Lapland during the winter 1882–83 an auroral station under S. TROMHOLT at Kautokeino, 69° N., 23° E.

France turned to the Antarctic regions for its choice of a location and decided on Cape Horn, where in the nearest safe harbor, Orange Bay, 55° 31′ S., 70° 21′ W., the station was installed, 6th September 1882, under charge of Lieutenant COURCELLE-SENEUIL. The entire expedition was under Captain F. MARTIAL, commanding the frigate *Romanche*, who observed the transit of Venus and explored hydrographically all of the Magellanic archipelago south of Terra del Fuego. The station was abandoned 3d September 1883, the expedition having experienced neither mishap nor disaster. Not only did they make observations, optional as well as obligatory, but brought back no less than 70 cases of specimens, in the domains of anthropology, botany, ethnography, geology and zoölogy. These collections have been classified, studied and discussed in a manner most creditable to French scientists, and the result given to the world in a series of eight illustrated quarto volumes, unequalled in their typographic beauty by any other of the International Polar publications.

Germany took a prominent part in the scientific work it had practically initiated, and established an Arctic and an Antarctic station, while the observatories at Breslau and Göttingen coöperated. The Arctic station was located at Kingawa Fiord, Cumberland Gulf, 66° 36′ N., 67° 19′ W., under Dr W. GIESE, who sailed from Germany, 28th June 1882, and returned 17th October 1883. Supplementary observations were made under Dr K. R. KOCH in Labrador, through the aid of the Moravian missionaries. GIESE carried out the complete programme of optional and obligatory observations, and in May 1883 explored adjacent regions, he sledging to the southeast, while L. AMBROERN examined the west side of the fiord.

The Antarctic station of Germany was on the island of South Georgia, where its members were landed by a ship of the German navy in September 1882. Under command of Dr K. SCHRADER the expedition quartered on the shores of Royal Bay, which is surrounded by enormous glaciers of a thousand feet, rising in the interior to 6,000 feet. These glaciers prevented extended explorations, and the mountain slopes were so steep that scarcely a day passed without the roar of heavy avalanches. While not strictly an Antarctic Island, yet its temperature conditions are such that the fauna and flora are wretchedly poor. On 3d September 1883, the party disembarked on a German gunboat, having successfully completed its obligatory and optional observations.

Great Britain and Canada coöperated in the establishment of a station at Fort Rae, Great Slave Lake, 62° 39′ N., 115° 44′ W. Captain H. P. DAWSON, leaving England 11th May, reached Fort Rae, 30th August 1882, and occupied the station to 1st September 1883.

Holland agreed to establish a station at Dicksonhavn, 73° 30′ N., 81° E., on the north coast of Asia. To this end Dr M. SNELLEN sailed in the Norwegian steamer *Varna* from Amsterdam, 5th July 1882. The ice-conditions to the westward of Nova Zembla were so unfavorable that the *Varna* entered the Kara Sea with great difficulty, and somewhat later, after an attempt to land on the west coast of Nova Zembla, was beset, 22d September 1882, in company with the Danish exploring steamer *Dijmphna*. From time to time ice threatened the destruction of the *Varna*, but she escaped serious injury until the end of December 1882, when the Dutch expedition sought shelter on the *Dijmphna*. Every effort to relieve the dangerous situation of the *Varna* failed, and 24th June 1883, she sank. Notwithstanding their unfortunate condition, Dr SNELLEN displayed the true scientific spirit by continuing regular observations from besetment to 1st August 1883, when the party started with boat and sledge to make the coast of Nova Zembla. They reached the south point of Waigat Island, 25th August, where they met the *Nordenskiöld* and were landed in Hammerfest, 1st September 1883—all in health.

The Norwegian station at Bossekop, 69° 56′ N., 23° E., under Assistant A. S. STEEN, was occupied from the middle of June 1882 to 31st August 1883.

The Russian stations were established under the auspices of the Imperial Russian Geographical Society. That in the Lena delta was located on Sagastyr Island, 73° 23′ N., 124° W., and occupied from 11th August 1882 to 6th July 1884. Lieutenant JURGENS, commanding, mapped carefully the whole delta, in addition to making complete observations, optional and obligatory.

The second Russian station, commanded by Lieutenant C. ANDREJEFF, was located on the west coast of Nova Zembla, at Little Karmakul Bay, 72° 23′ N., 52° 44′ E. While the expedition, numbering 13, was conducted under the auspices of the Imperial Geographical Society, yet the government contributed largely toward its equipment, and transported the party to Nova Zembla in the steamer *Tschishoof* of the Russian navy. Archangel was left 31st July 1882, and after an uneventful voyage they reached Karmakul, 4th August where their coming was welcomed by the bands of summer hunters, Samoyeds and Pomeranians, who in large numbers regularly visit Nova Zembla and occasionally winter on the southern island. That autumn the Samoyeds withdrew to the mainland, except one family that wintered near Karmakul. In May 1883, Dr GRINEWISKEY with two Samoyeds and a dog team crossed Nova Zembla to the Kara Sea, and later, with KRIWASCHEJA and a sailor, visited Matthew Strait, where geographic, botanical, and zoölogical specimens were obtained. On 1st December TISCHOFF, a sailor, left the house for the spring, a hundred yards distant: his absence was unnoticed, and the next morning he was found naked in the snow. When consciousness returned he could give no account of the circumstances, and after amputations he died. On 11th July 1883, the *Tschishoof* visited the station, and ANDREJEFF, leaving 5th September on the *Polar Star*, reached Archangel five days later.

The station designated for Sweden was Mussell Bay (NORDENSKIÖLD, 1872–73), near Grey Hook, the extreme northern point of Spitzbergen. Professor N. EKHOLM and eleven men sailed from Tromsöe, 9th July 1882, reaching Dane Island in six days, but the unfavorable ice-conditions obliged the ship to return southward. The station was then located in Ice Fiord, at Cape Thorsden, 78° 28′ N., 16° E. Here in 1872 a Swedish company in connection with an unsuccessful enterprise had built half a mile of railway and erected a habitation known as the Swedish House. It was, however, ill-omened from the fate of 17 Norwegians who perished in the winter of 1872–73, despite abundant fuel and food (p. 62). Nevertheless EKHOLM passed the winter without illness, disaster or unusual incident, and returned safely to Tromsöe on the gunboat *Urd*, 28th August 1883.

The United States were first in the field, through the indefatigable exertions of Captain H. W. HOWGATE, U. S. Army, whose *Florence* expedition to Cumberland Gulf in 1877 was followed by an unsuccessful attempt of the *Gulnare* in 1880 to reach Lady Franklin Bay.

Lieutenant A. W. Greely, who had declined command of the private venture, accepted command in 1881 when the expedition was made national by Act of Congress. The party consisted of four officers, 19 men of the army, including an astronomer, a photographer, and meteorologists especially enlisted for the purpose, and two Eskimo. Greely, in the sealer *Proteus*, left St Johns, Newfoundland, 7th July 1881, and after hunting up the mail of the British Arctic expedition, 1875–76, at Littleton Island, steamed through ice-free waters to Cape Lieber, 81° 37′ N. Here, on 3d August, in sight of its destination, the ship was stopped by the yet unbroken ice of Hall Basin; but she succeeded in entering Discovery Harbor 11th August, landed party and supplies at the winter quarters of the *Discovery*, 1875–76, and eight days later sailed homeward.

Quarters were speedily built and scientific work at once commenced. Freezing temperatures set in 29th August, to continue nine months, and the sun left 15th October, to be absent 135 days. Successful sledge-journeys broke the monotony of the autumn, and dry quarters, suitable food, hunting, exercise, and amusements insured perfect health through a winter unequalled for its severity.

Lieutenant J. B. Lockwood and Dr O. Pavy displayed great energy and endurance in preliminary sledging, even before the return of the sun in 1882. The ice-conditions of Robeson Channel were ascertained, Thank God Harbor (Hall, 1871) visited, and a small depot established at Cape Sumner. For six days Pavy's party travelled in temperatures averaging 73° below freezing, and Lockwood's for ten days in 74°.5 below,—periods of intense prolonged cold never encountered by any party in the field, either before or since. These extreme experiences gave confidence, since they were endured without injury to even the weakest, — and in Arctic work no team is stronger than its feeblest member.

Disease had killed two-thirds of the dogs, but two small teams still remained. The stronger was sent with Dr Pavy, who left 19th March 1882 to discover land to the north of Grinnell Land. Following the well-known coast to Cape Henry, he took to the polar pack, which was in the same indescribable roughness and confusion as Nares's officers found it in 1876. Four miles from land, 82° 56′ N., a gale housed the party, when to their dismay it was found, 23d April, that the storm had disintegrated the main polar pack, so that their floe was adrift in the Polar Ocean. Fortunately the floe set against the land near Cape Henry, and Dr Pavy escaped with such articles as were indispensable to immediate safety. The abandonment of the journey necessarily followed. The water seen by Pavy was local, for at the same time Lockwood, on the Greenland coast, was struggling over an ice-clad sea that seemed eternally bound, so thick were its floes.

Meanwhile the most important explorations were proceeding under Lockwood, who left Fort Conger 3d April with four two-man sledges, seven dog sledges, 13 men all told, under orders 'to explore the coast of Greenland near Cape Britannia,' which was suspected to be separated from Greenland, and in case he passed beyond that point, he was to proceed north, east, or 'in such direction as [he] thought best to carry out the objects of the [main] expedition, — the extension of knowledge regarding lands within the Arctic Circle.'

Depots at Cape Beechey and Newman Bay facilitated the operations, which up to the latter point were carried out under extreme cold (81° below freezing), rough ice and violent gales. From Newman Bay four men unfit for field work were sent back, and the advance party of nine started north 16th April with 300 rations for Repulse Harbor, which was reached overland in five days through tremendous exertions.

The journey onward was marked by severe storms, rough ice, broken sledges, snow-blindness, minor injuries, and — worst of all for loaded sledges — soft deep snow; nevertheless the party reached Cape Bryant, 27th April. The average daily travel to this point was nine miles, the greatest ever made by man-power in a very high latitude on any extended journey. It was within two and a half of the average attained 600 miles to the south, over ordinary ice, by the great Arctic sledgeman, McCLINTOCK.

LOCKWOOD remained two days at Cape Bryant, during which BRAINARD, RALSTON, and ELLISON visited the highest point of Cape Fulford, where a clear view confirmed the opinion of BEAUMONT, 1876, that St Andrew's Fiord probably unites with St George Fiord, and that no ice-cap covers the land northward of the 82d parallel.

Most reluctantly the supporting sledgemen, ELLISON, FREDERICK, LYNN, RALSTON, and SALOR, turned back on 29th May, when LOCKWOOD, BRAINARD, and Eskimo CHRISTIANSEN, with dog-sledge and 25 days' rations, struck northeast across the Polar Sea to Cape Britannia. Five and a half marches brought them to this cape; their travel was marked by deep soft snow, necessitating half-loads, and by the discovery of a water channel, with no bottom at 137 fathoms, separating the main sea-floe from the inshore ice.

From an elevation of 2,050 feet at Cape Britannia, LOCKWOOD saw that Victoria Inlet was 'an immense fiord running to the south; no land visible at the head; all to the south is an indistinct mass of snow-covered mountains,' ending in about 82° 30′ N. BRAINARD says: 'An occasional glacier of moderate dimensions could be seen struggling toward the sea,' the north limit of the glacial ice-cap of Greenland.

Having reached land never before trodden by man, they rounded Cape Frederick, crossed Nordenskiöld, Chipp, and Mascart Inlets, and camped 7th May at Low Point, 83° 07′ N., in equal latitude with the highest known land. Beyond Cape Ramsay, 83° 12′ N., the land ran east, and in twelve miles travel they lost two miles latitude, but at the De Long Fiord they turned north. The immense fiords of De Long, Chipp, Nordenskiöld, and Victoria, already passed, showed no signs of heading, and clearly indicated a new archipelago, intersected by these waterways. On 10th May, LOCKWOOD reached Mary Murray Island, 83° 19′ N., 42° 21′ W. A violent gale delayed them 63 hours, the cold exhausting them physically and the delay mentally. If weather forbade travel, life must be sustained; but they tasted insufficient food only at intervals of 15, 24, and 19 hours, — the last as clearing weather made progress possible. Floes so high that the sledge was lowered by dog-traces, ice so broken that the axe cleared the way, and widening water-cracks in increasing numbers impeded progress; but, despite all obstacles, they reached, 13th May 1882, Lockwood Island, 83° 24′ N., 42° 45′ W., the farthest of their journey, and the highest north, then or now.

On a clear day, from height of 2,600 feet, was seen to the north 'an unbroken expanse of ice, interrupted only by the horizon.' Their view extended far beyond the 84th parallel, and it was certain that the Polar Ocean there reached within 350 miles of the North Pole. To the northeast, in 83° 35′ N., 38° W., they saw the most northern known land, — Cape Washington, 28 miles to the north of Cape Columbia, Grinnell Land. There was a faint possibility of land extending northward from a point to the east of Cape Washington, but this was uncertain. To the south was 'a confused mass of snow-capped peaks, and the country much broken by entering fiords,' unfavorable to any extended ice-cap, as already indicated by the few small glaciers seen *en route*. Foxes, hares, lemmings, ptarmigan, and plants showed a country by no means devoid of vegetation or game.

Returning to the supporting party at Cape Sumner, Fort Conger was reached 1st June 1882.

In 1883, Lockwood attempted farther explorations to the northeast. Improved equipments and field experience insured more rapid travel, and in six days they travelled to Black Horn Cliffs, east of Repulse Harbor, — a journey that took 21 days in 1882. Arriving here 1st April, 24 days earlier than the previous year, open water confronted them. From an elevation of 1,300 feet, Lockwood saw water running northwest toward Cape Henry; the northern horizon was fog-covered, and the precipitous Black Horn Cliffs were washed by the open sea. The temperature fell to 73° below the freezing point, and the inshore water froze so as to bear a sledge; but on attempting to proceed, the entire pack set off shore to the north, leaving Lockwood adrift. Escaping with difficulty, he abandoned farther explorations, as his orders required.

While coastwise journeys were being made to the north, Greely penetrated Grinnell Land, with a two-man sledge. Starting 26th April 1882, and following Discovery Harbor to the southwest, a short overland journey took them to Chandler Fiord, which had been passed by the British expedition of 1875–76 as a small bay. Following it west, it proved to be a deep fiord 30 miles long, with a branch to the north near its western extremity. Proceeding up this branch, Greely came to a (Ruggles) river through which discharged a large glacial (Hazen) lake some 500 square miles in area, which at the junction of the two presented an ice-free pool. Crossing the lake, its feeding glaciers to the north proved to be projecting and escaping portions of a glacial ice-cap that covered North Grinnell Land between the Garfield Mountains and the Polar Sea. Henrietta Nesmith glacier had a convex front of some five miles, with a sheer perpendicular rise of 175 feet. Low, rounded and practically snow-bare hills formed the southern boundary of the lake, rising gradually to the high cliffs of Chandler Fiord.

This remarkable interior region was again visited by GREELY in June 1882, when he with two men reached Lake Hazen overland, and extended his discoveries to the southwest. Following up Very River to its source, the farthest reached was 175 miles from the home station, between Mount C. A. Arthur and Mount C. S. Smith, which evidently form the divide of Grinnell Land, — between Kennedy Channel to the east and the Polar Ocean to the west. GREELY ascended, 4th July 1882, Mount C. A. Arthur, the highest peak of Grinnell Land, 4,500 feet above the sea. The cold of altitude admitted of brief delay; but he saw that to the north of Lake Hazen were only snow-clad mountains, while the distant country to the west-southwest was similarly covered, — evidently ice-caps from their unbroken snow in very midsummer. The most distant mountains seemed to be on a separate land to the southwest, beyond an intervening fiord. The northwest view indicated that the United States Mountains, trending to the north, join the Challenger range of ALDRICH.

These discoveries were supplemented in 1883 by the journey of LOCKWOOD across Grinnell Land from Archer to Greely Fiord. This officer, with BRAINARD and CHRISTIANSEN his companions in his great northern explorations, left Fort Conger 24th April, and in four marches was 67 miles away, at Ella Bay, the farthest attained by ARCHER, 1876, in 14 days by man-sledge. Climbing the highest mountain, 4,400 feet, LOCKWOOD saw that the land to the south was covered by an ice-cap, of which the Ella Bay glacier is an offshoot. Persistently seeking another route, he penetrated by Beatrix Bay to Musk-ox Valley, — a musk-ox was there killed, — which was walled in by very high cliffs.

Reducing weight and food to the lowest limits compatible with his plan of twelve days' travel, LOCKWOOD pushed on with a small sledge taken for side trips. The country soon developed magnificent physical conditions, for the ice-capped region to the south presented to their astonished gaze a vertical ice-face ranging from 125 to 200 feet as measured by the sextant. Journeying west, they sought in vain for a defaced front in order to ascend it, but only two places presented slopes, neither practicable. As the glacier front ran across valleys and mountains with almost unvarying thickness, the temporary name of 'Chinese wall' was given to it, later changed to Mer-de-glace-Agassiz. At an elevation of 2,600 feet they crossed the 'divide' of Grinnell Land, and by a steep rocky ravine plunged down into a lake that debouched into the head of an unknown (Greely) fiord. A tidal crack, with its saline efflorescence, told that the western sea was reached. To the southwest they continued the magnificent shores of Greely Fiord, until driven to shelter by a prolonged storm that prevented farther progress. Their farthest camp was in 80° 48′ N., 78° 26′ W., whence from the adjoining cliffs, 2,200 feet high, the country was examined. Greely Fiord proved to be 10 to 15 miles wide and about 60 miles long. Its shores, great cliffs broken by occasional side fiords, end to the north abruptly in Cape Brainard, doubtless the west coast of Grinnell Land, and to the south in a succession of capes, the last Cape Lockwood, 60 miles distant, whence possibly the shore trends south to the vicinity of Schley Land, west of Hayes Sound. This journey ended the important geographic explorations, although unknown ground was elsewhere surveyed.

The discovery of Mer-de-glace-Agassiz placed the lake region of GREELY between two ice-caps of which the southern was the more remarkable, and the presence of Greely Fiord explained the manner by which the fleeing musk-ox and migrating Eskimo reached Grinnell Land and northern Greenland, from the Parry archipelago and the barren lands of North America.

The inland journeys of GREELY and LOCKWOOD resulted in the examination of about 6,000 square miles of newly discovered land, which determines satisfactorily the extent and the remarkable physical conditions of North Grinnell Land. It brought to light fertile valleys, supporting herds of musk-oxen, an extensive ice-cap, rivers of considerable size, and a glacial lake (Hazen) of extensive area. These valleys, 150 by 40 miles in extent, were 700 miles to the north of the point in Greenland where NORDENSKIÖLD sought similar physical conditions. Their discovery confirms his sagacity in believing in the possibility of such ice-free regions, which were forecast both for Greenland and Grinnell Land by Sir JOSEPH HOOKER, who thought them 'instead of ice-capped, merely ice-girt islands,' a prediction near the truth as regards Grinnell Land.

More remarkable, perhaps, was the discovery that Eskimo had wintered, as shown by permanent huts, at Lake Hazen,— doubtless a phase of that migration, remarkable for its route and distance over so barren a country, by which the children of the ice passed from the islands of the Parry archipelago to the east coast of Greenland. Lake Hazen, Black Rock Vale, Sun Bay, Discovery Harbor, Cape Baird, Cape Beechey,— all gave abundant traces of the journey across Grinnell Land.

Successful to such a degree as were these geographic explorations, they were strictly subordinated to the obligatory observations in the interests of the physical sciences. Systematic and unremitting magnetic observations served to round out knowledge by enabling scientists to calculate the secular variation of the magnetic declination of the Smith Sound region. Apart from the general value of the meteorological series, it has most fully determined the climatic conditions of Grinnell Land. The tidal observations were so complete at the station, and so amply supplemented by outlying stations, that scientists have determined not only the cotidal lines of the Polar Ocean with satisfactory results, but also learned from them that the diurnal inequality of the tidal wave conforms at Fort Conger to the sidereal day. The pendulum observations have been classed as 'far the best that have ever been made within the Arctic Circle,' and the 'determination of gravity [therefrom] has been singularly successful.' Botanical, zoölogical, and anthropological researches were pursued with similar unremitting attention, so that the scientific work of the expedition may be considered as satisfactory and complete, — especially in view of the high latitude of the station.

The visiting ship not coming, GREELY crossed to Kennedy Channel and found it ice-free. It appeared later that this ship, *Neptune*, remained inactive in Pandora Harbor during a southwest gale that there endangered her safety, while clearing the west coast and leaving open water to Fort Conger. Before the sun returned, in February 1883, GREELY laid down stores at Cape Baird to facilitate retreat by boats if the ship again failed. Field work over, arrangements were made in July to abandon the station;

records were duplicated, alternative articles selected, and everything packed for the worst. The ice breaking up 9th August, GREELY started the following day, carrying every pound of food and fuel possible in his steam launch, two boats and dingy, the last loaded with screened coal. From Fort Conger to Cape Hawks the voyage was marked by the trying experiences incidental to navigation through water-ways crowded with ice, acted on by strong currents and high winds. The heavy spring-tides, 25 feet or more, and recurring heavy gales, kept the heavy pack in constant motion, to and fro against the precipitous and rock-bound coast. Time and again only the most desperate efforts and measures secured the safety of the specially strengthened launch, while the whale-boats escaped destruction only by speedy unloading and drawing-up on floes. Every cache, however small, was taken up, ending with damaged, mouldy bread, etc., at Cape Hawks. This point was made in 16 days, the distance of 200 miles along the shore-line being doubled by the devious route necessitated by adverse conditions of ice and weather. Beset and frozen-in while attempting to cross from Hawks to Bache Island, the party waited ten days for the ice to break up. Abandoning the launch, the party struggled with desperation 19 days before it could reach the shore that was only 13 miles distant when they started. However, on 29th September a landing was effected half-way between Cape Sabine and Isabella, in health, with records, instruments and every essential or important belonging. But they were shelterless, shut in north and south by glaciers, and with small supply of food, despite constant seal-hunting during besetment. Some hunted on land, others on ice; some put up stone huts, others searched for cairns and records. The winter quarters were barely finished, when scouts brought news of the *Proteus*, which had been crushed north of Cape Sabine, 23d July 1883, by being forced into the ice by Lieutenant E. A. GARLINGTON, commanding relief party.

GARLINGTON's record ran in part: 'Depot landed . . . 500 rations of bread, tea and a lot of canned goods. Cache of 250 rations, left by expedition of 1882, visited by me, and found in good condition. English depot in damaged condition, not visited by me. Cache on Littleton Island; boat at Isabella. U. S. S. *Yantic* on her way to Littleton Island, with orders not to enter the ice. . . . I will endeavor to communicate with these vessels at once. Everything in the power of man will be done to rescue the [GREELY'S] brave men.'

It transpired that there was no boat at Isabella; that GARLINGTON's orders to replace damaged caches were imperative and disobeyed; that he had no knowledge that the Littleton Island cache was safe; that at Sabine he took every pound of food he could reach, though told that GREELY was provisioned only to August 1883; and that after COLWELL's skill had brought GARLINGTON safe to the *Yantic* he did not even ask WILDE to go north and lay down food for GREELY, otherwise doomed to starvation.

The drift experiences had demonstrated the impossibility of then crossing Smith Sound, and GREELY turned to Sabine, expecting the promised relief that autumn. Glaciers were rounded, hills scaled, new ice traversed, and the 15th October saw the party on Bedford Pim Island. Winter had begun, the polar night was imminent, clothing in rags, fuel wanting, and 40 days' rations must tide over 250 days, till help could come. The main party put up a hut of rocks, canvas, boat and snow-slabs, while selected men scoured the coasts for caches, sought land-game and watched seal-holes, until utter darkness drove all to the hut. Scientific observations were unremittingly made, amusements devised, a spring campaign planned, and the returning sun found only one dead. Efforts to cross Smith Sound failed, and a hunting-trip to the west found a new (Schley) land, but no game. Finally game came, so inadequately that food failed, and one by one men died, — JENS seal-hunting and RICE striving to bring in a cache. Courage and solidarity continued; and if GREELY gave to the maimed ELLISON double food while it lasted, he did not hesitate to order in writing the execution of a man, serving under an assumed name of HENRY, who repeatedly stole seal-skin thongs, the only remaining food. Flowers, plants, sea-weed and lichens eked out life for six, till 22d June 1884, when the relief ships *Thetis* and *Bear*, under Captain W. S. SCHLEY and Commander W. H. EMORY, rescued them. Records, instruments, and collections were saved to tell the story of an expedition that failed not in aught intrusted to it, and whose members perished through others.

The relief squadron of 1884, fitted out under the personal orders of the Hon. W. E. CHANDLER, Secretary of the Navy, passed the 'middle ice' with the first whalers, their rivals; and by their energy and daring rescued the remnant of the party during a violent gale. SCHLEY's cruise threw into sad relief the incompetency of the previous relief expeditions, but with him COLWELL showed again his qualities by first reaching the dying survivors.

The United States sent its second party to Point Barrow, the northernmost point of Alaska, where in 71° 16′ N., 158° 40′ W., Lieutenant P. H. RAY landed, 8th September 1881, with nine men. Friendly relations were cultivated and maintained with the Eskimo, 137 in all, who seek Point Barrow for sea game, their principal subsistence. With native aid RAY on several sledge-journeys made interesting geographic discoveries, reaching in the interior 69° 55′ N., 158° W., whence (Meade) mountains were seen to the south. He followed the east coast to 71° 01′ N., 154° 32′ W., discovering to the east of Dease Inlet a large river (Meade), which there empties by five mouths: the upper river is frequented by the Eskimo for the fish in its waters and the game in its stunted copses.

On 25th June 1882, the whaler *North Star* arrived; 13 days later, crushed by ice, she sank opposite the station, leaving her crew of 47 practically destitute. RAY promptly extended all needed assistance until, five days later, the whaler *Bowhead* arrived. Lieutenant J. W. POWELL visited the station 20th August 1882, in the *Leo*, bringing three men and taking away one. The last whaler passed south, 23d September 1882, and the first came north, 1st August 1883. RAY abandoned the station 27th August, in the *Leo*, having carried out the international programme with marked success. The natural history collections were large and representative, and the ethnographic studies and collections, made and discussed by MURDOCH, are interesting and important.

WILD: *Communications of the International Polar Commission* (St Petersburg 1882–84); TROMHOLT: *Under the Rays of the Aurora Borealis*, 2 v. (London 1885); LANMAN: *Farthest North; Life of J. B. Lockwood* (New York 1885); GREELY: *Three Years of Arctic Service*, 2 v. (New York 1886). Official Reports, see Chapter xviii.

CHAPTER XVII

GREENLAND

CONTINENTAL Greenland, with its outlying islands, extends from north to south more than 1,400 miles; and along the 78th parallel from east to west, Capes Alexander and Bismarck are 900 miles apart. Cape Farewell is seven degrees south of the Arctic circle, in the latitude of Stockholm, while Cape Washington is within six degrees of the North Pole.

Greenland is an elevated plateau, ranging in the main from 2,000 to 7,000 feet above sea level. Its precipitous, rocky coast is broken by numerous intersecting fiords, which are, or have been, beds of the extensive projecting glaciers that thus debouch from the parent inland-ice. This permanent ice-sheet covers nine-tenths of Greenland, and attains an unknown thickness, possibly 3,000 feet.

Three portions of the coast are inhabited. A band of Eskimo, the most northerly known inhabitants of the earth, live on the ice-free shores between Capes York and Alexander, entirely cut off by the surrounding ice-cap from communication with the outside world. They number some 270, and live very largely on sea game, although they were for many years without boats.

In East Greenland, between Capes Bismarck and Farewell, are scattered Eskimo tribes, numbering 600 souls or more, who reached their present habitat from the Parry archipelago by journeying around the north end of

Greenland through the strait (probably Nordenskiöld Inlet) that separates the continent from the new northerly land discovered by the Lady Franklin Bay expedition.

A few ruins mark the coming and passing of the first discoverers of South Greenland, who, almost pre-historic, were doubtless Norsemen; when and how they vanished are themes of endless discussion. That they were numerous, European, and Christian, appears from HOLM's investigations in 1880, among the ruins of the Julianehaab district, of which there are 100 known groups, with the structures in each numbering from 1 to 30. HOLM visited 40 groups and 300 separate ruins; among them were 4 well-built stone churches, one being 26 by 65 feet in size, and certain graves had crosses and carved figures therein. Lieutenant D. BRUNN's archæological explorations of 1894, in the same district, are likewise interesting and important.

The discoveries along the well-known west coast bordering Davis Strait and West Greenland Channel are elsewhere recorded.

The principal population of Greenland consists of the Danish Eskimo, 10,207, in 1892, who occupy the west coast from Cape Farewell as far north as Tasiusak, some 1,200 miles. They are governed by a strict monopoly, a bureau of the Danish government, the Royal Greenland Board of Trade, which well subserves the interests of the natives through its excellent corps of chief-traders and assistants, who are under the eye of two officers, inspectors, vested with magisterial powers and responsible to the crown.

East Greenland is but partly explored, largely owing to its inaccessibility. HUDSON, 21st June 1607, sighted land at the 73d parallel. On the authority of GERRIT VAN KEULEN's old chart, — good authority save that the latitudes are somewhat uncertain, — the following whalers visited this

No. XI.

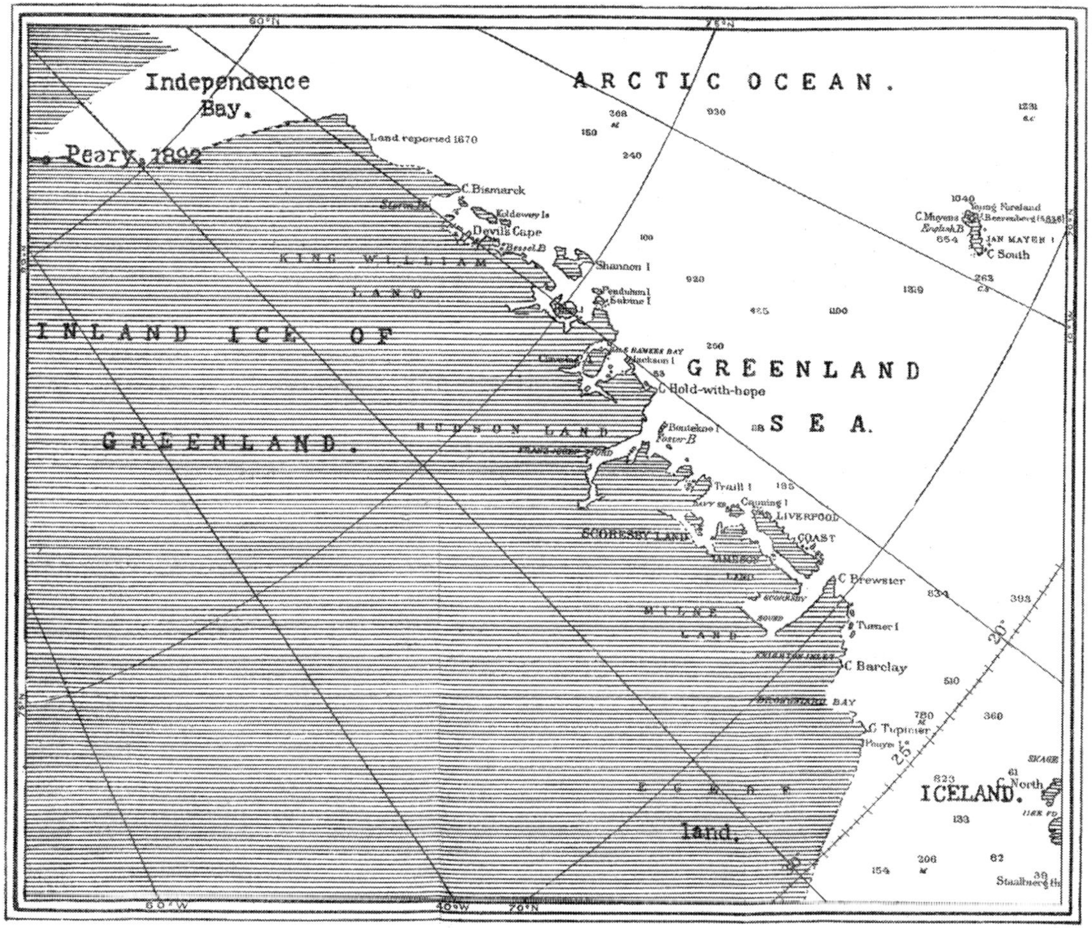

EAST GREENLAND AND JAN MAYEN (Hydrographic Chart, U.S. Navy)

coast during the seventeenth century: GALE HAMKE (1654), 74°.5 N.; BROER RUYS (1655), 73°.5 N.; EDAM (1655), 70° N.; and LAMBERT (1670), 78°.5 N.

The difficulty of approach to this coast, hemmed in by heavy ice and hidden by almost continuous fog, is shown by the fact that for more than 200 consecutive years its adjacent seas have been annually scoured by scores of daring and skilful whalers, who have made no material additions to the scanty and indefinite knowledge of the continental shore acquired in the 17th century. At long intervals a glimpse of land here and there, between the 73d and 80th parallels, served to make the geography of Greenland less mythical; but even at the beginning of the 19th century, the coast on the 75th parallel was laid down 14 degrees of longitude too far to the eastward.

Among expeditions that contributed to a knowledge of the East Greenland coast the following may be very briefly recited. The Dane DANNELL, in June 1652, skirted the ice-belt at a distance varying from 8 to 60 miles from 64° 50′ N., southward to the vicinity of Cape Farewell. HAN EGEDE, 1723, and OLSEN WALLØE, 1752, attempted, from the western settlements, to explore the coast by the use of Eskimo women-boats; the former barely passed the 60th parallel, and the latter was stopped at Nenese, 60° 56′ N. VOLQUARD BOON, a Dutch whaler, in 1761 discovered the bay named for him. EGEDE, the younger, in August 1786, and April–May, 1787, and LØWENORN in July 1787, followed the edge of the ice to the 66th parallel, but could not get within two miles of the coast. BLOSSEVILLE, in the French ship *Lilloise*, sighted land 29th July 1833, in 68° 34′ N., and traced it to 68° 55′ N. HAAKE, in July 1831, visited the coast near 74° N. The Scotch whaler, DAVID GRAY, in July 1868, discovered Scott's Inlet, 73° 30′ N. Between 7th and 11th July

1879, Captain WANDELL in the exploring Danish steamer *Ingolf* surveyed the coast from 66° N. to 69° N., but he could not approach nearer than six miles to the high mountainous and ice-clad land.

The first real and important discoveries in East Greenland are those made in 1822 by one of the best known of the famous Scotch whalers, Captain WILLIAM SCORESBY, jr., who with his father had seen the coast in 1817 and in 1821. On 8th June 1822, SCORESBY was sufficiently near the land in the *Baffin* to enable him to sketch and chart it from Hold-with-Hope, of HUDSON, 1607, in 73° 30′ N., to Gale Hamke Bay, 75° N., named for its discoverer in the Dutch whaler *Orangeborn*, 1654.

In intervals of fishing, during June to August, SCORESBY surveyed the land with great care and accuracy by astronomical and trigonometric observations on shore and at sea. The elder SCORESBY, in the *Fame*, explored Scoresby Sound and other inlets. The son was indefatigable in acquiring knowledge regarding the land and adjacent sea, which fortunately was given to the world in his *Voyage to the Northern Whale-Fishery*. The coast, barren, rugged and mountainous, had an average elevation of 3,000 feet, and rose occasionally to 6,000. Frequent fiords indented the coast, and the depth of Scoresby Sound, named after the father, led SCORESBY to erroneously believe that it bisected Greenland, — an opinion harmonizing with the existing theory about the country.

Prodigious numbers of birds, — little auks, plover, redpole, etc., were seen; butterflies, mosquitoes and bees appeared; horns of reindeer and bones of other animals were found. The recent presence of natives was indicated by ashes, sledge-runners, domestic implements, bone, iron-tipped arrows. Remains of the dead showed permanent residence, but one body in a wooden coffin evidently marked the passage of another race, — an ancient Norseman or a Dutch fisher.

Geographically SCORESBY'S discoveries were greater in importance and number than those of any other single navigator in East Greenland waters. It was not that alone he surveyed and charted with unusual accuracy a coast some 800 miles in its windings, but that he entirely changed the geographic features of Eastern Greenland. On former charts the coast between the 69th and 75th parallels was laid down with a southwest trend, covering 23 degrees of longitude from 5° to 28° W. SCORESBY reduced the longitudinal extent of East Greenland by nearly three-fourths, and determined the coast direction to be almost due north and south, between longitude 19° and 25° W. His scientific work, done in the intervals of most successful whale-fishing, was so abundant, comprehensive and intelligent, that it may safely be said that no other save NORDENSKIÖLD has contributed so materially to a scientific and accurate knowledge of the Arctic regions.

Captain EDWARD SABINE'S great pendulum work of 1823 comprised observations at the nearest possible points to the North Pole, and after visiting North Cape, Norway, and Hakluyt headland, Spitzbergen, with Captain CLAVERING, in the *Griper*, they turned their voyage to Greenland. SABINE landed at Pendulum Islands, 74° 32′ N., 18° 50′ W., in August 1823, and successfully carried out his scheme of pendulum work. CLAVERING skirted the coast by ship from Cape Parry, 72°.5 N., to Shannon Island, 75° 12′ N., whence the bold, mountainous shore was visible to the 76th parallel. By a boat journey of two weeks he also explored Gale Hamke Bay, and penetrated an ice-fiord where there were discharging glaciers of immense size. To his intense surprise CLAVERING found that this desolate coast was inhabited. Twelve

Eskimo occupied sealskin tents on the beach, and the presence of graves indicated permanency of residence. Unacquainted with other races the natives suddenly retreated after brief relations with CLAVERING's men. Doubtless these Eskimo reached East Greenland from the Parry archipelago, north of the continent of America, via Grinnell Land, Robeson Channel, northwest Greenland, Victoria, or Nordenskiöld, Inlet (of LOCKWOOD) and Independence Bay (of PEARY).

Southeast Greenland was first visited by Lieutenant W. A. GRAAH, Danish Navy, who sought diligently and unsuccessfully for traces of the 'lost colonies' of the *Easter Bygd.* Obliged to winter at Julianehaab, where he examined the ruins of that coast, — now thought to be those he sought, — he passed Cape Farewell in April 1829, and going northward explored the east coast to the vicinity of Cape Dan. His farthest point, Dannebrog Island, 65° 16′ N., was reached 18th August. GRAAH was the first one to chart with any approximate degree of accuracy the whole of the East Greenland coast from 60° N. to 73° N. GRAAH met many natives who had never seen whites, established friendly relations and wintered among them, 1829–30. As one result of this journey, traffic sprang up between the two coasts, and gradually the Eskimo are abandoning the more barren eastern coast. GRAAH's ethnographic and geographic work constituted the main stock of knowledge regarding southeastern Greenland for half a century, to the voyage of HOLM, 1880.

The next important contributions to knowledge are due to the second German North-polar expedition, which has been alluded to in Chapter xii. Captain KARL KOLDEWEY commanded the expedition in the ship *Germania*, with the *Hansa*, Captain HEGEMANN. The fortunes of the *Hansa* are elsewhere described, that unfortunate ship being separated from the *Germania*, through misunderstanding of a signal, was crushed by the ice, her crew escaping to Frederiksthaal, Greenland.

KOLDEWEY forced his way in the *Germania* through the ice-stream, and at a favorable point, on 5th August 1869, anchored at Pendulum Island, where the expedition wintered. Investigations showed that the Eskimo settlement of CLAVERING was abandoned, although remains disclosing long residence were found.

KOLDEWEY and PAYER explored that autumn Fligely Fiord and Kuhn Island, in a journey of 133 miles. Other neighboring fiords and lands were visited, and in the spring of 1870 by sledge KOLDEWEY and PAYER reached, 15th April 1870, 77° 01′ N., the highest point ever attained by explorers of the east coast, but which was exceeded by PEARY in his great ice-journey from West Greenland, when he reached Navy Cliff, 81° 38′ N., 34° W.

The most remarkable geographic discovery was made after the *Germania* was freed from the winter-ice. She entered Franz Josef Fiord, which penetrates inland far into Greenland, reaching five degrees of longitude to 73° 11′ N., 25° 58′ W. Its magnificence is set forth by PAYER as a combination of 'huge walls, deep erosion fissures, wild peaks, mighty crevassed glaciers, raging torrents and waterfalls.' One rock rises 5,600 feet out of the sea, and Mount Petermann is some 12,000 feet high.

In addition to these geographic contributions, the work of the German expedition resulted in an extremely important series of scientific papers on Arctic subjects. These valuable scientific contributions form Volume ii. of the German account of the expedition, and do not appear in the English translation.

Scientific explorations of value have been made by GIESECKE, HELLANDS, BROWN, NORDENSKIÖLD, and RINK.

The pioneer work of GIESECKE in geological investigations of Greenland is notable and important. For seven years, 1806–1814, he worked summer and winter pursuing his mineralogical researches, which covered the entire inhabited west coast, from 60° N. to 73° N., and for nearly a century his results have formed the general basis of geological publications relating to Greenland.

Especial interest has always attached to the question of the extent of the inland-ice, the name usually given to the ice-cap that covers the interior of all Greenland. This inland country has been pictured as consisting in part of extensive valleys, clothed with luxuriant vegetation which serves as pasturage for numerous reindeer, — an opinion that recent explorations have fully disproved. Occasionally ice-free areas, always barren and of limited extent, exist, which are called *nunataks* by the Eskimo, who believe them to be the dwelling-place of people (Kivigtoks or sorcerers) fled from human society.

These explorations fall naturally into two classes, the scientific and the adventurous. Several unsuccessful efforts have been made to cross Greenland from west to east, the most striking being that of Major PARS, in 1728, who attempted the journey with an armed mounted force. Efforts of DALAGER, RAE, BROWN, and WHYMPER to explore the ice-cap were likewise fruitless.

The route of DALAGER from Frederikshaab was followed again, 14th July to 5th August 1878, by Lieutenant J. A. D. JENSEN, a Danish officer, who penetrated to a distance of 47 miles, to 62° 50′ N., 48° 57′ W. The ice was here 5,000 feet above the sea, but above it projected a series of *nunataks*, or uprising ice-free peaks, to which JENSEN's name has been given.

In 1870 NORDENSKIÖLD attempted to penetrate Greenland from the head of Auleitsivik Fiord. Abandoning the sledges when the irregular ice made farther progress impracticable, the party proceeded on foot, carrying packs. The Eskimo finally refused to go farther, but NORDENSKIÖLD and BERGGREN went on until failing provisions necessitated their return. They were then over 35 miles inland, at an elevation of 2,200 feet above the sea, when they saw a deep broad river flowing rapidly between the blue banks of ice. Followed some distance, the whole stream plunged down a perpendicular cleft to an unknown depth. Two ravens seemed the only signs of life, but the acute eye of the botanist, BERGGREN, discerned a true ice-plant, a brown polycellular alga, while NORDENSKIÖLD discovered a gray powder, which he called kryokonite, a dust of cosmic origin. After this journey, the so-called meteorites were found at Mount Ovifak near Disco Fiord, and the greater number were brought to Sweden in 1871, the largest two weighing 20 and 9 tons respectively.

NORDENSKIÖLD visited Greenland again in 1883, and made 15 marches on the inland-ice from Sofia Harbor, at the head of Aulaitsivik Fiord, south of Disco Bay. The route followed closely the mythical fiord, that was supposed to bisect Greenland, as charted on EGEDE'S map, 1788. He camped 21st July in 48° 15′ W., at an elevation of 4,900 feet. Finding progress exceedingly slow, owing to crevasses and slopes, NORDENSKIÖLD sent his skilled Lapp *ski*-runners to go as far as they could and return in a few days. These men travelled 140 miles up gradually rising ice to 68° 32′ N., 42° 51′ W., where at an elevation of 6,600 feet the sea-cap still rose. In the light of HOLM'S explorations in East Greenland, it is now known that NORDENSKIÖLD thus reached by his Lapps a point nearer Egede or Sermilik Fiord, 66° 30′ N., 38° W., than to Sofia Harbor. NORDENSKIÖLD'S success, greater than before attained, first made it evident that Greenland could be crossed, and was the forerunner of NANSEN's crossing to the south and PEARY's to the north. NORDENSKIÖLD then visited East Greenland, and with difficulty made King Oscar Harbor, Cape Dan, 65° 35′ N., 37° 30′ W., 4th September 1883, beyond GRAAH's farthest. NORDENSKIÖLD's hydrographic researches in the Spitzbergen Sea were extensive and important.

The first crossing of Greenland was made over the inland-ice by Dr FRIDTJOF NANSEN, who decided that the most practicable method was to be left on the east coast, when necessity would insure success. With five others, NANSEN, after six weeks' fruitless efforts to reach the Greenland coast in a Norwegian sealer, attempted to land, 17th July 1888, by boat at Cape Dan, 63° 20′ N., from which they were separated by an ice-stream ten miles wide. The two days set apart for forcing the pack passed, and they were still in the sea, beset and sweeping southward. It was not until 29th July, after bitter suffering and strenuous effort, that they made Anoritok, 62° 05′ N., near 200 miles to the southward of their contemplated landing. Following the shore northward and meeting natives, they commenced their ice-journey from Umivik, 64° 45′ N., which they reached 10th August.

In addition to the usual Alpine outfit, Canadian and Norwegian snowshoes, they carried instruments, food, fuel, sleeping-gear, and other supplies, a load of 1,200 pounds for their five sledges. Steep, irregular slopes, soft snow and dangerous crevasses so delayed progress, that by 27th August, they were only 40 miles inland in 64° 50′ N., at an elevation of 7,000 feet. The lateness of the season constrained NANSEN to shorten his journey and change his direction toward Godthaab, 64° N., instead of Christianshaab, 68° N., his original destination.

A broad flat plateau, between 8,000 and 9,000 feet high, forms the crest of southern Greenland, the inclination to the west being gentler than the eastern ascent. Long distances were sailed over rapidly until the dangerous crevasses prevalent near the coast necessitated caution, and the rough irregular ice retarded progress. A journey of 260 miles brought them, 29th September, to Kangersunek Fiord, 50 miles south of Godthaab, whence assistance was obtained by means of a boat improvised of canvas, willows, and sledges.

The crossing of Greenland by NANSEN is justly regarded as a brilliant feat, in which the boldness of the plan was matched by the energy and endurance that overcame the physical difficulties. The journey proves that the same conditions of unbroken ice-cap obtain over extreme southern Greenland, as were disclosed by the explorations of NORDENSKIÖLD some 200 miles farther north.

Denmark has done an enormous amount of exploration and investigation of the edges of the inland-ice, of its debouching glaciers, of its barren *nunataks*, of the fauna and flora of the ice-free coasts, which, supplemented by observations of climatic and hydrographic character, make Greenland a well-known country to the 75th parallel on the west coast, and to the 66th on the east coast. On the west coast alone the area of the inland-ice region thus explored covers about 950 miles from north to south, with an average width of 24 miles, being about 2,250 square miles. This work has been done in a most thorough manner, a definite section being assigned each year to specialists suited to the work in hand.

RINK, the best authority on Greenland, gives the following summary of the fields covered, to include 1886:

K. J. V. STEENSTRUP (naturalist), with G. HOLM (naval officer), and KORNERUP (naturalist), traversed the district of Julianehaab, (60°–61° N.) in 1876; in 1877, with A. JENSEN (naval officer), the Frederikshaab district (61°–62° 30′ N.). In 1878, STEENSTRUP journeyed to North Greenland, where he spent two winters. In the same year JENSEN, with KORNERUP and GROTH (draughtsman), travelled over the Frederikshaab and Godthaab districts (62°–64° N). In 1879, JENSEN with KORNERUP, and R. HAMMER (naval officer), inspected the coast from 67° to 68½°, after which HAMMER passed the winter there. The summer of 1880 found STEENSTRUP and HAMMER in North Greenland, while G. HOLM, GROTH and PETERSEN (naturalists) continued the exploration of the southern promontory of Greenland. In 1881, HOLM, with SYLOW (naturalist), visited again the country around Cape Farewell. In 1883, HAMMER, SYLOW, and LARSEN (naval officer) traversed the west coast from 67° to 70° N. HOLM, W. GARDE (naval officer), and KNUTSEN and EBERLIN (naturalist) undertook the journey to the east coast, which lasted till 1885, and necessitated the establishment of winter quarters twice. Meanwhile JENSEN, in 1884, followed the west coast from 65¼° to 67° N., with RUSS-CARSTENSEN (painter), while, at the same time, a cruise of the man-of-war *Fylla* was utilized for scientific purposes. In 1885, JENSEN, RYDER (naval officer) and S. HANSEN (anthropologist) traversed this coast from 64½° to 65½° N. Lastly, in 1886, RYDER BLOCK (naval officer), and USSING (naturalist) went to Upernivik (72½°–75° N.), where the first two spent the winter, and the *Fylla* visited Davis Strait.

These important explorations of both the western and eastern coasts pertain especially to the ice-fiords and bordering inland-ice, and are fully treated in *Meddelelser om Grønland.* (See Chapter xviii.)

The most important geographic work was that of HOLM on the east coast, 1883–85. Lieutenant HOLM wintered on west coast of Greenland, 1883–84, and on 5th May 1884, put to sea with seven hunters in kayaks, and his party of six transported in skin boats, rowed by 19 women and five men. Cape Farewell was rounded, and the dangerous and tedious journey northward along the east coast began. Often delayed by a close ice-pack crowded against the precipitious and glacier-faced coast, the expeditionary force passed the time as best they could, while the natives, especially the women, indulged in eating, drinking, smoking and merry-making. In Lindenows Fiord, 62° 15′, were impenetrable willow groves, while near this site of the only Scandinavian ruins on the east coast the shore was lined with driftwood. This coast was traversed by HOLM and KNUTSEN to 66° 08′ N. Five ice-fiords were discovered, and according to the reports of the natives there was another in 68° N. Icebergs in extraordinary numbers lined the coast, the largest about 260 feet high. Between 60° and 63° N., GARDE found nearly 200 glaciers reaching to the sea, of which 70 were a mile broad.

In 1883, at the farthest reached, Kasingortok, provisions had been stored under charge of an Eskimo, NAVFALIK. They were found untouched in 1884. What honesty and self-denial these natives exhibited may be judged from the statement that their utmost endeavors barely furnish food for maintaining life.

At Aneretok, they met natives who had once visited the west coast for trade purposes; they had long, sleek, black hair, wonderfully-oval and European-like faces, but some of the women had fair hair. Seventeen days were consumed by delays at Karoakornak, between Capes Adelai and Ranzau, the point where GRAAH had such bad fortune in 1830. On 18th July, 19 of the Danish Eskimo insisted on returning. Proceeding to Tuigmiarmint Fiord, HOLM sent GARDE and EBERLIN back to the west coast, which they safely reached in September, and himself wintered at Angmagsalik, 65° 37′ N., in 1884–85. The natives at this point had never seen Europeans before. There were eleven communities, aggregating 431 souls. There were 548 natives south of the 68th parallel, and CLAVERING, as before mentioned, saw 11 in 74° 05′ N.

In 1891, an important Danish expedition under Lieutenant RYDER visited Northeastern Greenland. In July he reached the vicinity of Pendulum Islands, and landing at Cape Hold-with-Hope, wintered in Scoresby Sound, at Hekla Harbor, 70° 27′ N., 26° 12′ W. The inner fiords were explored 50 miles from the coast, so it is evident that these fiords extend inland farther than was supposed. Unfavorable ice-conditions prevented RYDER from getting south of 69°. Reaching Iceland 20th August 1892, RYDER again revisited the East Greenland coast and landed, 10th September 1892, at Tasiussak, visited by NORDENSKIÖLD in 1883. Thence RYDER went north and visited Tasiusarsik, where HOLM wintered 1884–85, and reached the most northerly inhabited point, Minatikii, Sermiligak Fiord, 15th September. From the mountain peak of Islet Ananak, it was seen that the mainland was ice-clad. The natives in 1892 occupied all settlements and numbered 293 persons, a reduction of 221 since 1885; 114 of them had gone south, and 107 were dead. RYDER saw 32 kinds of birds, 160 flowering plants, a few fishes, bears, foxes, reindeer, musk-oxen, hares, lemmings and seals.

Dr DRYGALSKI, of the Berlin Geographical Society, wintered on Nugsuak peninsula 1892–93, examined the glacier deposits and collected fossils; he also measured the movement of the Karayak ice-stream. He believes that the motion and work of the ice-cap depends on the action of the solids and fluids composing it, particularly of the enormous quantity of water which permeates it.

Charged with the survey of the Julianehaab district in 1893, Lieutenant T. V. GARDE ascended the inland-ice near Sermiatsialik fiord and crossed to Ikersuak. About 100 miles inland, in 61° 54′ N., the crest was passed at an elevation of 7,300 feet, the ice thence sloping to the east and northeast.

The same summer R. KNUDSEN approached Blosseville Land, and saw no signs of the inland-ice at any place. The foreground between 29° 12′ and 39° 50′ W. was low, but between Cape Grivel and Nuna Isua, 68° N., where the land, projecting more decidedly to the south than appears by the Danish map of 1888, was very high and occasionally precipitous.

The last and most important work on the East Greenland coast has been the establishment by the Danish government of a missionary, trade, and meteorological station at Angmagsalik, and the closing of that coast to other nations. This ensures the future welfare of these natives, under the same beneficial methods that have marked Danish sway in western Greenland. Captain HOLM landed at Angmagsalik fiord 26th August 1894, and constructed for these purposes two buildings on the shores of King Oscar Harbor.

The most brilliant work on the inland-ice is that of Mr R. E. PEARY, U. S. Navy, who with a Dane, MAIGAARD, reached a point some 50 miles from the sea, near Disco, in 1886. Renewing his explorations in the *Kite*, PEARY landed at M'Cormick Bay, August 1891, and most courageously persisted in his work, although his leg was broken while crossing Melville Bay. A house was erected, but autumnal efforts to establish a cache at Humboldt glacier were futile. In 1892 PEARY, able to travel, explored Inglefield Gulf in April, and then turned to the accumulation of stores at the edge of the inland-ice, some 15 miles distant. His main journey commenced 14th May, when the true inland-ice was reached, with 16 dogs and 4 sledges. He crossed the divide of 5,000 feet elevation, between Whale Sound and Kane Sea, and at a point 130 miles from M'Cormick Bay sent back COOK, who had supported him thus far with a man and two dog-sledges. PEARY proceeded with ASTRUP, and looked down into Petermann Fiord, 31st May, but crevasses here and at St George Fiord obliged them to make a detour to the east and southeast. Finally, on 26th May, they reached the north edge of the inland-ice near 82° N., whence they looked to the north on the brown-red, comparatively ice-free land discovered by LOCKWOOD in 1882. The fiord, into which they could not descend, doubtless connects with Nordenskiöld Inlet of LOCKWOOD, 1882, and PEARY supports GREELY'S opinion of 1884, that Greenland here ends, and that the discovery of LOCKWOOD is an entirely new land.

Unable to go farther north, PEARY turned to the southeast to make the east coast of Greenland; and following the edge of the ice-cap, reached Independence Bay, 4th July 1892, and climbed Navy Cliff, 4,000 feet high, 81° 37′ N., 34° W. To the north was an ice-free land extending to the east some 50 miles, to 25° W. longitude; to the east and southeast, the East Greenland ocean was covered by disintegrating sea-ice. Five musk-oxen were killed, which relieved anxiety for dog-food on the homeward trip. The return journey to M'Cormick Bay, about 450 miles distant, was made almost in a straight line, the ice-divide proving to be 8,000 feet above the sea.

Believing that even more extended discoveries could be made in northeast Greenland by again crossing its ice-cap, PEARY, raising funds for the purpose by a series of lectures, established a station at Bowdoin Bay in 1893. With 8 men, 12 sledges, and 92 dogs, he ascended the inland-ice

6th March 1894, and in 13 days advanced 134 miles to an elevation of 5,500 feet. Stormbound by violent gales and extreme cold, PEARY saw his dogs die and his men frosted, so that a general advance was impossible. Caching all surplus stores, principally pemmican, he sent back the disabled force, and with indomitable, but fruitless energy, marched on with three selected men. In 14 days he travelled only 85 miles, under extremely adverse conditions that finally obliged him to return with dying dogs and failing men. Abandoning sledges and caching pemmican he reached Bowdoin Bay on 15th April, with only 26 living dogs of the original 92.

Later ASTRUP, his chief support, sledged to Melville Bay and charted a considerable portion of its indefinitely located northeastern shore.

When the visiting steamer *Falcon* arrived, in August 1894, prudence demanded that the entire party should return to the United States. Food and fuel were insufficient, more extended explorations were improbable, and arrangements for a visiting ship in 1895 were merely problematical. With determination and courage bordering on rashness, PEARY decided to winter at Bowdoin Bay, with two volunteers, LEE and HENSON.

Utilizing throughout the winter the entire resources of the region and gaining Eskimo recruits, PEARY accumulated supplies on the inland-ice, and started northward 2d April 1895, with his 2 men, 4 Eskimo, and 63 dogs drawing 6 sledges. The third march an Eskimo deserted with his outfit, but PEARY, undiscouraged, pushed on. Most unfortunately the heavy snows had obliterated all landmarks, and the expected mainstay — the pemmican cache — could not be found. Failure now impended, but sending back his Eskimo allies, from this camp, 134 miles inland and 5,500 feet above the sea, PEARY continued his journey, 41 dogs dragging the 3 sledges. The temperatures ran from −10° to −43°, the elevation increased to 8,000 feet, travel was bad, sledges broke down, LEE was frosted, dogs died; but PEARY persisted on his hopeless journey. Finally, with but 11 exhausted dogs, 1 sledge, and a disabled man, PEARY 8th May left LEE camped 16 miles from the coast, and with HENSON sought game ahead unsuccessfully for 4 days. Scant walrus-meat reserved could barely feed their dogs during the home journey; but with desperate courage they advanced their camp to Independence Bay, PEARY'S farthest in 1892. The descent to the sea practically destroyed their sledging equipment; but 10 musk-oxen restored vigor to men and dogs. Farther game failing, with 9 dogs and food for 17 days they turned homeward in a frantic race against starvation. Twenty-five forced marches, in which necessarily everything but food was abandoned, brought them in desperate condition, 25th June, to Bowdoin Bay, whence by the steamer *Kite* they reached Newfoundland 21st September 1895.

If PEARY'S advance beyond his buried cache was one of the rashest of Arctic journeys, yet the courage, fertility of resource, and physical endurance displayed by him and his companions place their efforts among the most notable in Arctic sledging. Other parties under less desperate circumstances have met with mortality, and only escaped total fatality by relief from their reserve party, which adjunct to Arctic exploration experience indicates to be essential to safety.

The two crossings of Greenland by PEARY must be classed among the most brilliant geographic feats of late years, his journeys far surpassing in extent that of his ice-cap predecessor, NANSEN, who crossed Greenland more than 1,000 miles to the south. PEARY'S efforts extended northward the east coast of Greenland, more than two degrees of latitude; and the increasing longitude of 15 degrees of the coast at Independence Bay tends to prove that LOCKWOOD'S new land to the north of Greenland is of limited extent, as has been advanced by several geographers.

The physical collections and observations enlarge the previously existing wealth of scientific data of western Greenland. Doubtless the most important scientific results derived from the Peary voyages are those connected with Professor CHAMBERLAIN'S examination of the glaciers of Inglefield Gulf, in which survey photography was freely used, and to great advantage. Geology must profit from this study of glaciers presenting such varied forms, especially as the unusually free exposure of structure facilitated examination of vertical faces, convoluted and laminated formations.

The most attractive additions to knowledge are the ethnological studies of the Cape York Eskimo, which, when published, should scarcely be inferior in interest to the very valuable ethnological memoirs of HOLM on the Eskimo of East Greenland.

EGEDE: *Description of Greenland* (London 1745); CRANTZ: *History of Greenland,* 2 v. (London 1820); GRAAH: *Expedition to East Greenland* (London 1837); ETZEL: *Grönland* (Stuttgart 1860); RINK: *Tales of the Eskimo* (London 1875); also, *Danish Greenland* (London 1877); JOHNSTRUP: *Giesecke's Mineralogiske Rejse i Grønland* (Copenhagen 1878); NORDENSKIÖLD: *Grönland* (Leipzig 1886); NANSEN: *First Crossing of Greenland,* 2 v. (London 1886); HOLM: *Øst Grønland Expedition,* 2 v. (Copenhagen 1888–89).

CHAPTER XVIII

BIBLIOGRAPHY

AT the conclusion of each chapter have already been given lists of the more important works that may interest the general reader. Here will be very briefly considered sources of detailed information.

The only attempt to compile a complete bibliography of the vast stores of Arctic literature is that of Dr J. CHAVANNE, aided by Drs KARPF and MOMMIER, published under the auspices of the Royal Geographic Society of Vienna. The arrangement of this work, *Die Literatur über die Polar-Regionen der Erde* (Vienna 1878), is topical and geographic, and in its 6,617 titles are also included circumnavigations and general works of travel. While it leaves much to be desired in fulness of detail, yet it is of decided value to all students of Arctic literature.

The most comprehensive body of data relative to scientific polar work is that prepared by the Arctic committee of the Royal Society for the British expedition of 1875, and published by the Admiralty as a *Manual of the Natural History, Geology, and Physics of Greenland and adjacent regions*, 92 and 795 pp., 3 maps (London 1875).

The Parliamentary Papers and Blue Books, which contain reports and journals of nearly every British Arctic expedition, 1803–1876, comprise a vast amount of information (unfortunately not indexed) relative to the Northwest Passage, Franklin Search, and explorations to the north of America. BROWN, in *Northwest Passage*, gives 15 titles of *Parliamentary Papers* and *Blue Books* pertaining to the Franklin Search alone.

The most important of these books are as follows: —

Blue Book: Arctic Expedition (RICHARDSON and RAE; KELLETT, *Herald;* PULLEN, *Plover;* SAUNDERS, *North Star;* J. C. ROSS, *Enterprise* and *Investigator*), 7th March 1850, 157 pp., map.

Blue Books: (1, 2 and 3); Report, journals, and papers connected with investigation of Austin and Penny expeditions: (1) Report, 25th November 1851, 199 pp., 2 maps; (2) Additional Papers, 22d October 1851, 368 pp.; (3) Farther Correspondence (containing also, De Haven U. S. Expedition; RAE, Wollaston Land; JOHN ROSS, *Felix;* PULLEN, Boat Journey; HOOPER'S Proceedings; COLLINSON, *Enterprise;* Proceedings, *Plover* and *Daedalus*), 216 pp., 3 maps, 1852.

Blue Book: Arctic Expeditions (KENNEDY, *Prince Albert;* INGLEFIELD, *Isabel;* MOORE, *Plover*), 20th December 1852, 88 pp., 2 maps (n. d.).

Blue Book: Papers (M'CLURE, *Investigator;* BELCHER'S squadron; KELLETT, *Resolute;* PULLEN, *North Star;* MAGUIRE, *Plover;* M'CORMICK'S Boat Journey), 225 pp., 5 maps, 1854.

Blue Book: Farther Papers, January 1855 (INGLEFIELD, *Phœnix;* BELCHER, *Assistance;* KELLETT, *Resolute;* M'CLURE, *Investigator;* TROLLOPE, *Rattlesnake;* MAGUIRE, *Plover*, including SIMPSON'S Western Esquimaux; COLLINSON, *Enterprise;* PULLEN, *North Star;* RAE, Repulse Bay, first traces of FRANKLIN'S fate; Sledge journals of BELCHER'S squadron), 958 pp., 26 maps, 1855.

Later have been published, *Blue Book:* Journals, etc.,

(NARES) Expedition, 1875-76, 484 pp., 20 maps (1877); C. 1636; continuation of C. 1153, 1875; and C. 1560, 1877. Report of Committee on Scurvy in the late (NARES) Arctic Expedition, London, 1877, lv, 505 p.

Of special lists, by far the most complete and satisfactory is that of publications and charts relating to Alaska and adjacent regions, by W. H. DALL and MARCUS BAKER, in the *Pacific Coast Pilot, Alaska*, vol. i, U. S. Coast and Geodetic Survey (Washington 1879). It comprises 3,832 titles and sub-titles in eleven languages, and is simply invaluable to any student of the Alaskan regions.

As regards Greenland, the following works are extremely important: *Grønlands historiske Mindesmerker*, 3 v. (Copenhagen 1838–45); *Meddelelser om Grønland*, 16 v. (Copenhagen 1879–1895). The Danish text of the latter work is in most volumes supplemented by a summary in French, while vol. xi, RINK, *Eskimo Tribes*, is in English. Vol. xiii, LAURIDSEN: *Bibliographia Groenlandica*, is especially useful with its 2857 titles, rich in scientific works, which include almost every publication relating to the country. The principal contents are: Vol. i, — JENSEN, KORNERUP, LANGE, and HOFFMEYER, *Inland-ice of Godthaab and Frederikshaab Districts*, 1878; ii, — STEENSTRUP, KORNERUP, JENSEN, G. HOLM, and LORENZEN, *Geological Investigations Julianehaab*, 1876, and of *Egedesminde and Holsteneberg*, 1879; iii, — J. LANGE, C. JENSEN, BRANTH, GRØNLUND, ROSTROP, and KOLDERUP, ROSENOMGE, *Conspectus Floræ Groenlandicæ, 4 parts*; iv, — HAMMER, STEENSTRUP, and LORENZEN, *Glacial, Geological, and Geographical Investigations in Jacobshavn, Rittenbenk, Umanak, and Upernivik Districts*, 1878–80; v, — STEENSTRUP, O. HEER, and DE LORIOL, *Cretaceous and Miocene Flora of West Greenland*, 1878–80; v (Supplement), — O. HEER, *Greenland's Fossil Flora*; vi, — WANDEL, NOR-

MAN, and G. HOLM, *Greenland East Coast and Julianehaab Ruins*, 1880–81; vii, LORENZEN, RØRDAM, WANDEL, LUNDBECK, and S. HANSEN, *Mineralogy of Greenland, Hydrography Davis Strait, Entomology and Anthropology of West Greenland, etc.*; viii, — HAMMER, JENSEN, RYDER, LANGE, WARMING, TH. HOLM, RØRDAM, RINK and CARLHEIM GYLLENSKIØLD, *Investigations in Disco Bay, Holstenborg, Sukkertoppen, Godthaab, and Upernivik Districts*, 1883–87; ix–x,—G. HOLM, V. GARDE, KNUTSEN, EBERLIN, STEENSTRUP, S. HANSEN, LANGE, RINK, VILLAUME–JANTZEN, and CRONE, *East Greenland Investigations*, 1883–85; xi, — RINK, *Eskimo Tribes*; xii, — E. WARMING, *Greenland Vegetation*; xiii, — LAURIDSEN, *Bibliographica Groenlandica*; xiv and xv, not seen; xvi, — V. GARDE *et al*, *Expedition to Julianehaab District*, 1893, etc.

The voluminous literature of Swedish explorations in Spitzbergen and Nova Zembla is shown by 202 titles in LESLIE: *Voyages of Nordenskiöld* (New York 1879).

Admiral A. H. MARKHAM has rendered a service to students in his *Bibliography of Hudson Bay* (not seen).

In BROWN: *The Northwest Passage*, 2d ed. (London 1860), are 273 titles, the larger number pertaining to the Franklin Search.

GREELY, in vol. ii *Lady Franklin Bay Expedition* (Washington 1888), gives 314 titles bearing very largely on meteorology and allied sciences.

Among lists of Arctic books published by public libraries, those of Boston and Providence, May 1884, are among the fullest.

Of Arctic scientific works, the following are only a few out of the many: RICHARDSON & SWAINSON: *Fauna Boreali-Americana*, 4 v. (London 1836); NORDENSKIÖLD *et al.*: *Studien und Forschungen* (Leipzig 1885), and *Wissenschäftlichen Ergebnisse der Vega-Expedition* (Leipzig, n. d.); *Wissenschäftlichen Ergebnisse 2d. Deutsche Nordpolarfarht*, 1869–70 (Leipzig 1874); *Smithsonian Contributions to Knowledge* (Washington): *Observations in the Arctic Seas*, — No. xi, KANE'S Meteorological; No. xiii, M'CLINTOCK'S Meteorological; No. xv, HAYES'S Physical; GAIMARD: *Voyage en Islande et au Groenland*, 1835–36 (Paris 1838–51), 7 vols. and 4 atlases; GAIMARD: *Voyages en Scandanavie, en Laponie* (au Spitzberg 1838–40, Paris 1843–48), 16 vols. and 6 atlases.

Among the most important and extended individual publications relating to Arctic matters are those of Professor O. HEER. Based on Arctic fossils, principally from Greenland, Spitzbergen, and Nova Zembla, they tell the story of enormous climatic changes, outline the development of the vegetable kingdom and its distribution over the earth's surface. The principal titles are *Flora Fossila Arctica*, 7 v. (Zurich 1868–80); *Flora Fossila Gronlandica*, 2 v. (Zurich 1882–83).

The series entitled International Polar Scientific Publications constitute by far the greatest collection of scientific Arctic data and related memoirs extant. They number 31 quarto volumes, devoted almost entirely to observations and their discussions. As these publications are not well known, their titles are given: WOHLGEMUTH: *Osterreichische Polarstation Jan Mayen*, 3 v. (Wien 1886); PAULSEN: *Expedition Danoise, Godthaab*, 2 v. (Copenhagen 1889–93); DAWSON: *Fort Rae* (Great Britain) (London 1886); LEMSTRÖM & BIESE: *Expedition Polaire Finlandaise*, 3 v. (Helsingfors 1886–87); HYADES, LEPHAY, CANNELLIER, *et al.*: *Mission Scientifique du Cap Horn*, 8 v. (Paris 1885–91); NEUMAYER & BORGEN: *Beobachtungs-Ergebnisse der Deutsche Stationen*, 2 v.; I., *Kingua Fjord*; II., *Süd-Georgien* (Berlin 1886); SNELLEN & VOLCK: *De Nederlandsche Pool-Expeditie*

(Utrecht 1886); STEEN: *Der Norwegischen Polarstation Bossekop in Alten* (Christiania 1888); ANDREJEFF & LENZ: *Beobachtungen der Russischen Polarstation auf Nowaja Semla*, 2 v. (St Petersburg 1886–91); LENZ & EIGNER: *Beobachtungen der Russischen Polarstation an der Lenamunding*, 2 v. (St Petersburg 1886–95); EKHOLM *et al.*: *Observations par l'Expedition Suedoise, Cap Thorsden*, 2 v. (Stockholm 1885–89, 1891–94); GREELY, SCHOTT *et al.*: *Proceedings of United States Expedition to Lady Franklin Bay*, 2 v. (Washington 1888); RAY, SCHOTT *et al.*: *Report of* [United States] *Expedition to Point Barrow* (Washington 1885); MURDOCH: *Ethnographic Results of the Point Barrow Expedition* (Washington 1893).

CO-OPERATIVE OBSERVATIONS

SOLANDER: *Observations du Magnetisme à Upsala* (Stockholm 1893).

INDEX

Printed in Dunstable, United Kingdom

91220009R00167